PINEAPPLES, PENGUINS, & PAGODAS

Traveling Around The World Through Literature, Research, & Thinking Skills

by Barbara Jinkins

Incentive Publications, Inc.
Nashville, Tennessee

*To
my husband Dennis
and daughters
Jennifer, Beverly, and Allison*

*Illustrated by Marta Johnson
Cover by Becky Rüegger
Some illustrations by Jennifer Jinkins
Edited by Jan Keeling*

ISBN 0-86530-258-8

Table of Contents

INTRODUCTION

What do the youth of today know about world geography? Studies indicate that most students lack a broad foundation of knowledge and awareness of the world around them. The purpose of this book is to introduce students to the seven continents of the world—the land regions, animals and plants, and the people of each. To encourage an awareness of geography, the five themes of geography, as described in the National GENIP (Geographic Education National Implementation Project) Program, are emphasized. These five themes are:

1. Location—position of the earth's surface
2. Place—physical and human characteristics
3. Relationships within places—humans and environments
4. Movement—people, goods, and ideas
5. Regions—how they form and change

This book is designed to be an invaluable addition to your "Whole Social Studies" classroom, with activities that strengthen research techniques, map skills, creative writing, and critical thinking. Also included are literature-based activities and enrichment activities in the

areas of science, math, art, music, and physical education. Although the activities are assigned to specific continents, most can be adapted to other continents to meet specific needs.

The tasks require students to utilize reference sources such as nonfiction books, magazines, encyclopedias, dictionaries, almanacs, and atlases. The students are encouraged to locate resources by using the card catalog. A strong cooperation between the librarian and the classroom teacher will reinforce the development of the required skills needed to successfully complete the activities.

Teachers are encouraged to highlight one of the seven continents each month. However, the activities can be used independently. Another method would be to emphasize specific skills by featuring certain activities from several continents. The contents of this book can be used for one week, one month, or one school year, depending on the objectives of the teacher.

TRAVELING AROUND THE WORLD

It is important for students to develop a broad exposure to the various cultures and land regions of the world. To accomplish this goal, numerous activities can be incorporated into all subject areas of the curriculum, including language arts, literature, social studies, science, math, art, music, and physical education. A combination of independent study, class discussions, and cooperative learning projects will further enhance the learning process. Many activities, especially those for the first continent to be studied, should be prepared in advance. As the class prepares for future "travels," invite the students to participate in the planning stages.

Before making your study plans, reproduce the letter on page 10. Send the letter home with each child to inform the parents of your "worldwide trip." As the letters are returned, compile a list of the resources available for you to use during your "travels."

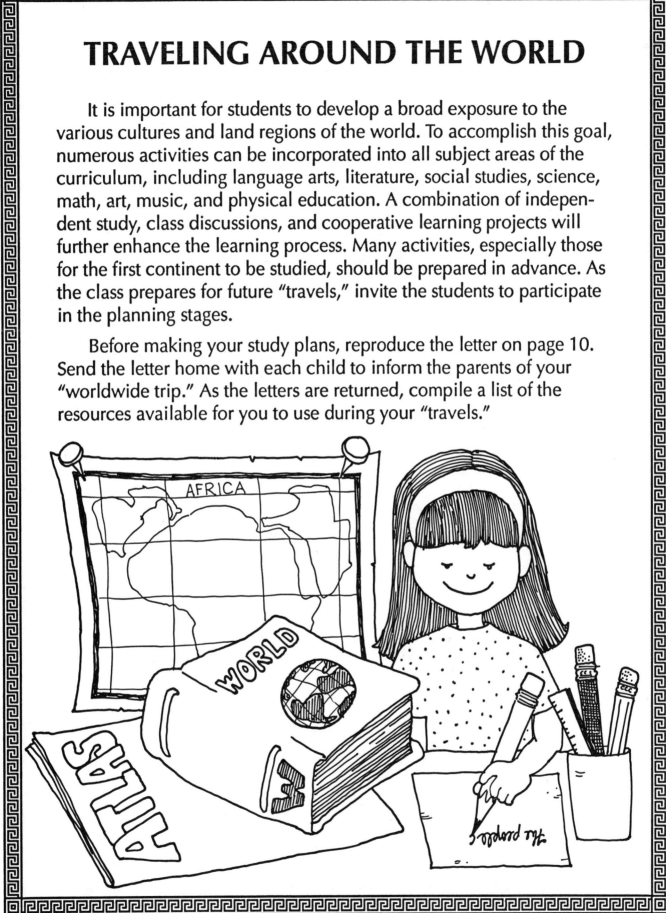

AMERICAS

Dear Parents:

Our class is preparing to "visit" various countries around the world during the school year. We are developing a list of resource people who can help us better understand the different regions and ways of life in other areas of the world.

Please complete the information below and return it with your child to the school as soon as possible.

Thank you for your interest and cooperation.

Sincerely,

_____

NAME OF PARENT(S): _____

PHONE NUMBER: _____

COUNTRIES I HAVE VISITED: _____

_____

_____

SOUVENIRS I WILL SHARE: _____

_____

_____

I ALSO HAVE: (please check)
_____ SLIDES
_____ FILMS / VIDEOS
_____ PICTURES

NAMES OF OTHER RESOURCE PEOPLE TO CONTACT:

_____ PHONE NUMBER: _____

_____ PHONE NUMBER: _____

AFRICA AUSTRALIA

EUROPE

ASIA

BEGINNING YOUR TRAVELS

1. Display a large world map in a prominent location. Mark on the map the countries where each student's family originated.

2. Begin your world tour by giving each student a "passport" and a "suitcase." The passport contains a list of activities designed to acquaint students with each continent. **Each activity is fully explained in the corresponding continent sections of this book.**

The suitcase will be used to carry each student's passport and completed activities. Directions for passport and suitcase are found on page 15.

Although a one-month study of each continent is suggested, the time frame for the tour may vary. The teacher should determine whether all or part of the activities for each continent should be completed by the students. This will depend on the time available and specific instructional emphases.

3. Decorate the classroom to represent the continent. If several classes are participating in the study, decorate each room to represent a specific land region.

AFRICA: vines, wild animals (pictures or stuffed), pyramids, waterfalls, huts, trees (carpet tubes can be used for the trunks)

ANTARCTICA: ice blocks (covered boxes), penguins, whales, icebreakers, research station

ASIA: lanterns, junk (made from a large box—great for quiet reading), nomad tent, kites, oriental rugs, pagoda, vines

AUSTRALIA: sheep ranch, eucalyptus trees, native animals, coral reef, rain forest

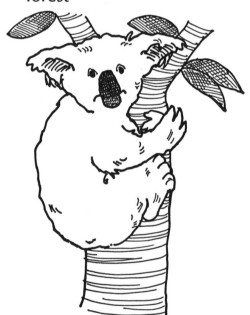

EUROPE: snow-capped mountains, Black Forest region (artificial evergreen trees), gondola (made from a large refrigerator box turned on its side), palm trees (for the Mediterranean region), enlargements of Big Ben and the Eiffel Tower, posters

NORTH AMERICA: maple trees, Canadian Mounted Police, mountains, totem poles, Mexican market, native animals, pineapple plantation

SOUTH AMERICA: cattle ranch, mountains, coffee crop, vines, Amazon River, snakes, native animals, masks

4. Ask travel agencies and airlines for posters and brochures. The class can write thank-you notes to the companies for their donations.

5. Call or write foreign embassies for information or to enlist guest speakers.

6. Contact the science and social sciences departments of local universities to solicit speakers.

7. Visit flag companies to borrow, rent, or purchase flags of various countries. The students can make many of the simple flags from construction paper.

8. Ask the school cafeteria manager about the possibility of offering a foreign menu in the lunchroom during your "travels."

FOOD SUGGESTIONS:

AFRICA: corn, peanuts, citrus fruits, grapes, bananas

ANTARCTICA: fish sticks, snow cones

ASIA: eggrolls, rice, vegetables

AUSTRALIA: meat pies, pavlova, pikelets (pancakes)

EUROPE: spaghetti, pizza, sausage, French bread

NORTH AMERICA: hamburgers, apple pie, tacos, refried beans

SOUTH AMERICA: tortillas, beef empañadas, arroz con leche (rice pudding)

9. Select videos or films that will provide examples of life in different countries.

10. Encourage your students to be "Globe Trotters." Each day during the week, give the class clues to a particular location on the continent being studied. Instruct the students to put their responses, along with the date and time of their answer, in an answer box. Do not announce the winner until Friday. If no one has identified the site by Friday, review the clues with the class and determine the location together.

EXAMPLE: BRAZIL

MONDAY: This is a South American country that borders the Atlantic Ocean.

TUESDAY: Most of the people here live along the coastline because rivers, forests, and mountains restrict inland travel.

WEDNESDAY: Thirty percent of the world's total coffee crop is grown in this country, which is sometimes called the world's largest "coffeepot."

THURSDAY: The official language of this country is Portuguese.

FRIDAY: This country's capital city is located at 15.47 degrees latitude and 47.55 degrees longitude.

11. To introduce a unit of study, allow each student to choose a question about the continent from a container. The student must answer the question and share his or her answer with the class before the end of the study. Establish a specific time for the students to present their information. If you would like each student to concentrate his or her research on a particular country, place in the container the names of nations located on the continent.

Examples of containers:

AFRICA: clay pot or urn

ANTARCTICA: fur-lined boot

ASIA: lacquered box

AUSTRALIA: palm leaf basket, backpack, or kangaroo pouch

EUROPE: candy box, flower pot, or European-made toy

NORTH AMERICA: piñata, bushel basket

SOUTH AMERICA: woven basket

Passport And Suitcase

Begin your world tour by giving each student a passport and a suitcase to decorate.

Passport

The cover design for the passport is at the bottom of this page. To save paper and copying time, you may copy this design twice on one piece of paper, then use that piece of paper to make the rest of the copies. Cut the covers apart and fold on the dotted line. Then make copies of the inside passport pages, found on pages 17–20. You will notice that each page is numbered at the bottom so you will be able to place them in the proper order. You will find that if you desire to save paper and create a sleeker passport, the pages can be arranged so that they can be copied on front and back sides of paper. Cut pages apart on solid lines, fold on dotted lines, place in order in passport covers, and staple together.

Suitcase

The "suitcase" provided for each student is simply a legal-size expandable file folder with added handles. Each student will use a suitcase to carry passport and completed activities. The pattern for the suitcase handles is on page 16.

WORLD TRAVELER

PASSPORT

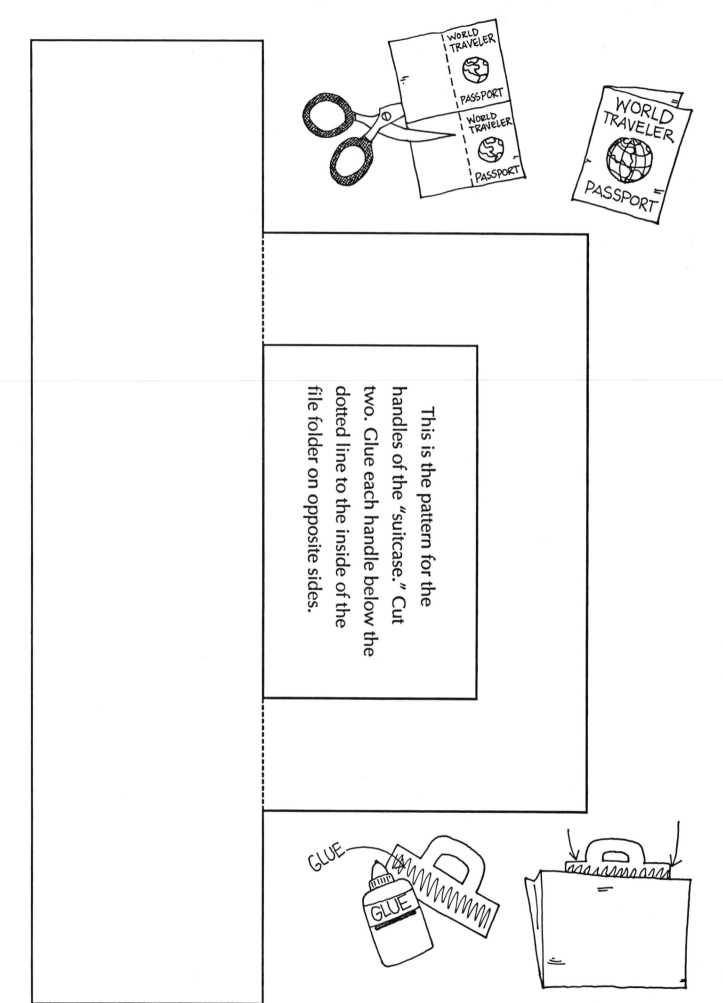

This is the pattern for the handles of the "suitcase." Cut two. Glue each handle below the dotted line to the inside of the file folder on opposite sides.

GLUE

16

ASIA

___ 1. Compare and contrast musical instruments.

___ 2. Illustrate an economy cube.

___ 3. Complete a crossword puzzle.

___ 4. Research kites.

___ 5. Describe an Asian landmark.

___ 6. Write a Japanese poem.

10

NORTH AMERICA

___ 1. Dress a heritage character.

___ 2. Identify some Canadian wildlife.

___ 3. Determine the best route for a boat trip.

___ 4. Plan a fiesta.

___ 5. Prepare for a hurricane.

7

AFRICA

___ 1. Construct an animal sack puppet.

___ 2. Discover valuable minerals.

___ 3. Examine the types of homes found in African regions.

___ 4. Produce a travel poster.

___ 5. Create a transportation diorama.

___ 6. Read a mystery book.

12

TEACHER AFFIDAVIT

I, _____ ,

a teacher at _____

_____

School, do hereby affirm that

_____ ,

the bearer of this passport, is a student of said school and is entitled to travel to the seven continents of the world during a period not to exceed the nine months that constitute a school year.

Signed_____

Date _____

Bearer_____

5

SOUTH AMERICA

___ 1. Make a gelatin box puppet.

___ 2. Compile a "shape" booklet of facts.

___ 3. Write a postcard to a friend.

___ 4. Design a billboard.

___ 5. Develop a product map.

___ 6. Read an animal book.

8

EUROPE

___ 1. Produce a shoebox filmstrip.

___ 2. Design a commemorative stamp.

___ 3. Measure the distance between two landmarks.

___ 4. Select European souvenirs.

___ 5. Read a folktale or fairy tale.

9

ENDORSEMENTS AND LIMITATIONS

This passport is valid for all countries that occupy the seven continents.

IMMUNIZATIONS:
All students are considered appropriately immunized since they have been officially accepted as students of said school.

ENTRY TO FOREIGN COUNTRIES:
This passport must be examined by customs and immigration officials when crossing borders and used as means of an official identification.

6

AUSTRALIA

___ 1. Contrast the continents of Australia and Antarctica.

___ 2. Investigate sports events.

___ 3. Reconstruct the Great Barrier Reef.

___ 4. Send a message in a bottle.

___ 5. Photograph a book.

___ 6. Create rock paintings.

11

DESCRIPTION OF BEARER

Name: _____

Homeroom No.: _____

Address: _____

Place of birth: _____

Hair color: _____

Eye color: _____

Issued: _____

Expires: _____

4

ANTARCTICA

___ 1. Decide how to protect this environment.

___ 2. Construct a coathanger face.

___ 3. Assemble an animal mobile.

___ 4. Organize a survival kit.

___ 5. Compose a news story.

___ 6. Read a biography.

13

PASSPORT

This passport is the property of the herein named student. Any misuse or alteration of this official document is strictly prohibited. This passport is valid only if signed by the bearer.

_____
(Official signature of bearer)

2

TRAVEL NOTES

15

TRAVEL NOTES

14

PHOTOGRAPH OF BEARER

THUMB PRINTS

Left Thumb   Right Thumb

3

The staff of _____ school requests that you allow the bearer of this passport to pass freely and without hindrance as he or she travels around the world. Allow bearer to spread the goodwill and friendship of our school to every country.

1

DURING YOUR TRAVELS . . .

The activities listed in this section are intended to supplement and enrich the study of each continent. All of the main curriculum areas are included.

LANGUAGE ARTS

1. Have foreign language dictionaries available. Students enjoy learning foreign terms for color words, numbers, names for family members, names of fruits and vegetables, and other common nouns. Let the class make dictionaries, language wheels, or flash cards to share. Instruct the students to write stories using their spelling words. They may then circle all of the nouns and try to find the corresponding words in a foreign language dictionary. If possible, invite foreign language students from a high school to share different languages with the class.

2. Write letters to different organizations requesting information:

Trans-Antarctic Education Programs
P.O. Box 4097
St. Paul, MN 55104
(You will receive information about Antarctica and receive the newsletter, THINK SOUTH.)

Australian Tourism Commission
3550 Wilshire Blvd.
Suite 1740
Los Angeles, California 90010

Ocean Alliance
Fort Mason Center, Building E
San Francisco, CA 94123
(You will receive information on how to adopt a whale)

Hershey Food Corporation
Park Boulevard
Hershey, Pennsylvania 17033
(You will receive a poster on the history of chocolate and the process of making it.)

Del Monte Corporation
San Francisco, CA 94105
(Ask for information about the areas where their crops of vegetables and fruits are grown.)

Folgers Coffee Company
A Subsidiary of Procter and
   Gamble
Cincinnati, Ohio 45202
(Ask where their coffee beans are grown and how they are processed.)

3. Gather the students on a "magic carpet." Use literature to share stories about various cultures. Select books from both the fiction and nonfiction

sections of the library. The librarian can assist you in locating appropriate selections.

Incorporate specific reading skills into the lesson. Have a globe available to locate the settings of the stories. At the end of each section you will find fiction books to share with your class. These selections relate to the different regions of the world and include activities for enrichment.

4. Invite storytellers to share stories from other lands.

5. Correspond with pen pals around the world.

6. List words associated with the different countries being studied. Merge them into the weekly spelling list.

7. Encourage the students to write imaginary stories that include facts about a specific country.

SOCIAL STUDIES

1. Instruct the students to display newspaper articles about various countries on a current events bulletin board. Discuss the information with the class.

2. Keep a class scrapbook of information and pictures about different countries. Encourage the students to research topics that interest them. (See "More Topics to Research" on the "Outline of Activities" page at the beginning of each section.)

3. Collect postcards from various countries.

4. Use old NATIONAL GEOGRAPHIC magazines to make puzzles of landmarks, lifestyles, or animals of the different continents.

5. Gather stamps from different countries and explore the significance of the picture on each.

6. Encourage the students to visit an ethnic grocery store. List items normally not found in an American store.

7. Exhibit ethnic newspapers for the students to examine. Let them discover what the stories are about by studying the pictures and comparing them to an American newspaper.

8. Prepare a foreign tasting party. Do not forget to emphasize unusual fruits or vegetables.

9. Research a famous person from each continent. Share an important contribution of that person with the class.

10. Design a travel brochure after reading about a specific country.

11. Prepare an area to exhibit items of interest. Try to gather a large variety of objects relating to the economy, current events, or lifestyles of different countries. Examples: wool to represent sheep farming in Australia, a piece of the Berlin Wall, a kimono from Japan, etc.

SCIENCE

1. Research inventions and inventors from other countries.

2. Track a major storm such as a hurricane.

3. Using the daily newspaper, graph temperatures of various cities of the world during a particular month. Discuss why they are different or similar.

4. Plant bulbs and record their growth.

5. Discuss natural disasters such as earthquakes, tornadoes, or volcanic eruptions. Using the WORLD ALMANAC, find out where such disasters have occurred.

6. Find ways to protect the natural environment. List reasons for erosion, pollution, endangerment of species, and the depletion of the ozone layer. Next, discuss ways to prevent these occurrences.

7. Invite local zoo representatives or pet store owners to bring unusual animals such as a llama, python, or tropical birds to your class. Discuss their habitats and countries of origin.

MATH

1. Measure the distances in both miles and kilometers between cities or landmarks. Learn to convert one unit of measurement to the other.

2. Invite the owner of an ethnic bakery or restaurant to visit your class to discuss his or her business and country.

a. Cook an ethnic recipe using either standard or metric measurements.

b. Divide the class into groups. Instruct the groups to order from a menu and total the bill. Don't forget to add the gratuity. (Great for teaching percentage!)

3. Take the students to a grocery store. Instruct them to list the items they find which are imported from foreign countries. Make a graph showing the different foods of the countries represented.

4. Learn about the different time zones. As a class, calculate the times at different locations around the world.

5. Create a Mexican market place. Price the items in pesos and learn to convert the amount to American dollars.

6. Invite a local banker to visit your class and bring foreign coins. Learn about converting American currency to foreign currency.

7. Study Roman numerals. Find out how they are still used.

ART

1. Build clay or salt relief models of the continents or a specific country.

2. Design a piñata for the class, or make individual ones from paper cups or lunch sacks.

3. Create a class totem pole using a carpet tube or a stack of boxes. Individual totem poles can be made from tissue tubes.

4. Tie dye t-shirts.

5. Learn to write your name using calligraphy.

6. Many cultures wear masks for special ceremonies and festivals. Design a mask.

7. Use paper sacks to make the costumes of different countries.

8. Make masks of native animals.

9. Use hieroglyphics to design pictures.

10. Weave cloth on a loom.

11. Learn to make baskets.

12. Mold and paint a piece of pottery.

13. Create a piece of batik art.

MUSIC

1. Learn songs from different countries.

2. Assemble a variety of musical instruments. List how they are alike and different. Learn to play at least one.

3. Invite dance groups to teach dances from various cultures.

4. Survey your community to locate people who play ethnic instruments such as bagpipes. Invite them to visit your class.

5. Read about different composers. Mark their home countries on a map.

PHYSICAL EDUCATION

1. Learn to play games from other countries.

2. List popular sports played across the world.

3. Invite athletes from various countries to speak to your class.

4. The sport of ballooning is popular in France. Provide a hot air balloon demonstration or film.

5. Research the most popular sports of different countries. Choose one or more sports and explain the rules to the class.

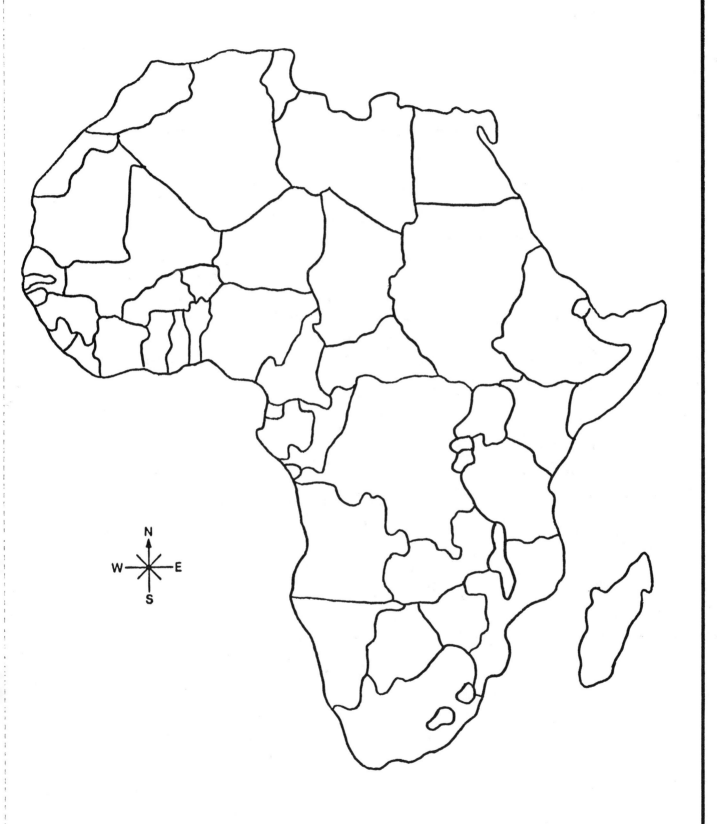

AFRICA | OVERVIEW

Africa, the second largest continent, is located south of Europe with the Atlantic Ocean on its west and the Indian Ocean on its east. It consists of 51 independent countries and is a land of great contrasts. The land area is covered by deserts, grasslands, and tropical rain forests. Over 800 languages are spoken by the people; Arabic, Swahili, and Hausa are the most common of these. About two thirds of Africa's population live in rural areas. These people raise livestock and grow crops mainly for their own needs. The continent has great mineral deposits of gold, petroleum, diamonds, and copper. Disease, over-population, famine, and extreme poverty remain immense problems for Africa, despite the great mineral wealth.

Northern Africa is mostly desert. The Sahara, the largest desert in the world, and the Nile, the longest river, are located in this region. Crops are grown along the coastlines. Most of the people in the north are Arabs. Nomads herd camels, goats, and sheep, looking for grazing land. This area is also the site of pyramids that are over 4,000 years old.

Western Africa is an area of rapid modernization, but the people remain very poor. The people who live south of the Sahara make up 800 ethnic groups. Grasslands and forests cover the land. Crops of cocoa, coffee, pineapples, bananas, cotton, peanuts, and rubber are grown for export.

Eastern Africa consists of grassland areas. Many wild animals live in this region. Crops of tea and coffee are grown here for export. Livestock is also raised. Desert areas can be found in the countries of Ethiopia, Kenya, Somalia, and Djibouti.

Southern Africa is mainly a large, flat plateau. Grasslands and the Nambi Desert occupy most of this land area, where some of the richest gold and diamond mines in the world are located. Great waterfalls prevent ships from using the rivers. However, the power from the falls can be used to generate electricity.

AFRICA COUNTRIES

INDEPENDENT

Algeria
Angola
Benin
Botswana
Burkina Faso
Burundi
Cameroon
Cape Verde
Central African Republic
Chad
Comoros
Congo
Djibouti
Egypt (African)
Equatorial Guinea
Ethiopia
Gabon
Gambia
Ghana
Guinea
Guinea-Bissau
Ivory Coast
Kenya
Lesotho
Liberia
Libya
Madagascar
Malawi
Mali

Mauritania
Mauritius
Morocco
Mozambique
Niger
Nigeria
Rwanda
Sao Tome and Principe
Senegal
Seychelles
Sierra Leone
Somalia
South Africa
Sudan
Swaziland
Tanzania
Togo
Tunisia
Uganda
Zaire
Zambia
Zimbabwe

DEPENDENT

Madeira Islands
Namibia
Reunion
St. Helena Island Group
Western Sahara

OUTLINE OF ACTIVITIES | AFRICA

1. **WILDLIFE SAFARI** *(p. 34)*—Select one African animal. Construct a paper sack puppet of that animal. After reading about the animal in the encyclopedia, list five interesting facts about it on the back of the puppet.

2. **THE WEALTH OF THE LAND** *(p. 35)*—Many valuable minerals can be found in Africa. The countries of Algeria, Libya, and Nigeria are major producers of petroleum, sometimes called "black gold," while South Africa ranks first in the world in the mining of gold. Discover more facts about these valuable minerals.

3. **HOMES IN AFRICAN REGIONS** *(p. 36)*—When one thinks of Africa, three distinct regions come to mind. These are the deserts, the grasslands, and the tropical rain forests. Research these regions. Draw the types of shelters in which the people who live in the rural areas of each region would dwell.

4. **SIGHTS TO SEE, PLACES TO GO** *(p. 38)*—Using the information that you read about the regions of Africa (#3), design a travel poster to advertise one of the regions.

5. **TRAVELING THROUGH AFRICA** *(p. 39)*—Transportation in Africa differs from region to region. Select a region. Using a shoebox, make a diorama to show one of the ways the people in that region travel.

6. **SOLVE A MYSTERY** *(p. 40)*—How would you nave solved the problem differently?

MORE TOPICS TO RESEARCH

Sahara Desert	Victoria Falls	Pygmies
Irrigation	Mt. Kilimanjaro	Petroleum production
Oasis	Bronze	Louis Leakey
The Great Pyramids	Ivory	Hieroglyphics
Sand Dunes	Mosque	Rosetta Stone
Diamonds	Cassava	Mummies
Gold	Nomads	Lungfish
Suez Canal	The Great Sphinx	
Baobab tree	Gnus	

Africa is known for its wild animals. Thousands of birds, animals, reptiles, and amphibians live in the grasslands, forests, and swamps across the continent. People are the chief enemies of these animals. The natural habitats of many animals are destroyed as cities are built. Other animals are overhunted. Several countries have formed game reserves and national parks to protect the animals.

Select an animal that lives in Africa. Construct a paper sack puppet of it. After reading about the animal, list five interesting facts about it on the back of the puppet.

FACTS ABOUT THE

_____

1. _____
   _____
   _____

2. _____
   _____
   _____

3. _____
   _____
   _____

4. _____
   _____
   _____

5. _____
   _____
   _____

THE WEALTH OF THE LAND | AFRICA

Many valuable minerals can be found in Africa. These include gold, petroleum, copper, diamonds, and uranium. The countries of Algeria, Libya, and Nigeria are major producers of petroleum, sometimes called "black gold," while South Africa ranks first in the world in the mining of gold. Find out more about these minerals by completing the chart below.

Leading gold-producing countries

1. _____
2. _____
3. _____
4. _____
5. _____

Uses for gold

1. _____
2. _____
3. _____
4. _____
5. _____

Leading petroleum-producing countries

1. _____
2. _____
3. _____
4. _____
5. _____

Uses for petroleum

1. _____
2. _____
3. _____
4. _____
5. _____

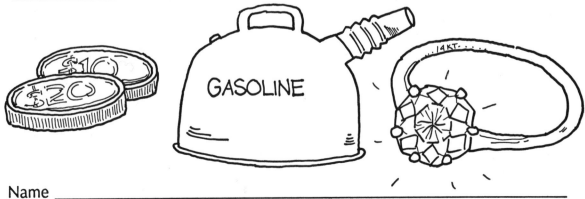

Name _____

The continent of Africa can be divided into three main land regions. These are the deserts, the grasslands, and the tropical rain forests.

Because the climates in each region are so different, the houses of the people also must be different. The people who live in the large cities usually dwell in modern types of houses, while those in rural areas live in the traditional houses of their villages.

After you have completed the information about the houses in each region, choose one of the regions and construct a village of houses. Be sure to include plant and animal life.

THE DESERT REGION

There are three deserts that cover approximately two fifths of the African continent. These are the Sahara, which is the largest desert in the world, the Nambi Desert, and the Kalahari Desert.

Sudan and Morocco are two African countries that are located in the desert region. Locate information about one of these countries and draw a picture of a rural house on a separate sheet of paper. Why is it best for the houses in this region to have thick walls?

THE GRASSLAND REGION

The grassland region, also called the savanna, covers approximately two fifths of Africa. Much farming takes place in this region. Nomads also herd sheep, goats, and cattle in this area, searching for water and land to graze.

Two grassland countries are Mali and Kenya. Read about these countries. On a separate sheet of paper, draw a rural house and a nomad's shelter. Name three ways these houses differ from each other.

THE TROPICAL RAIN FOREST

Many people believe that Africa is covered by dense jungles. You will find only a few true jungles on this continent. However, forests cover approximately one fifth of Africa. Most of these are tropical rain forests.

The Congo, Cameroon, and Gabon are three of the countries that are located in the tropical rain forest region of Africa. These countries receive 80–100 inches of rain each year. After exploring information about one of these countries, draw a typical rural dwelling. Why do you think people who live deep in the tropical rain forest would not choose wood as the main building material for their homes?

Africa, with its various land regions and diverse cultures, is a favorite place for tourists to visit. From sailing on the Mediterranean Sea to exploring the tropical rain forests, there are many sights to see and activities to enjoy.

Use the information that you gather about the regions of Africa to design a travel poster to advertise an activity or special place to visit. Interest others in your vacation spot by hanging the poster in your classroom.

TRAVELING THROUGH AFRICA 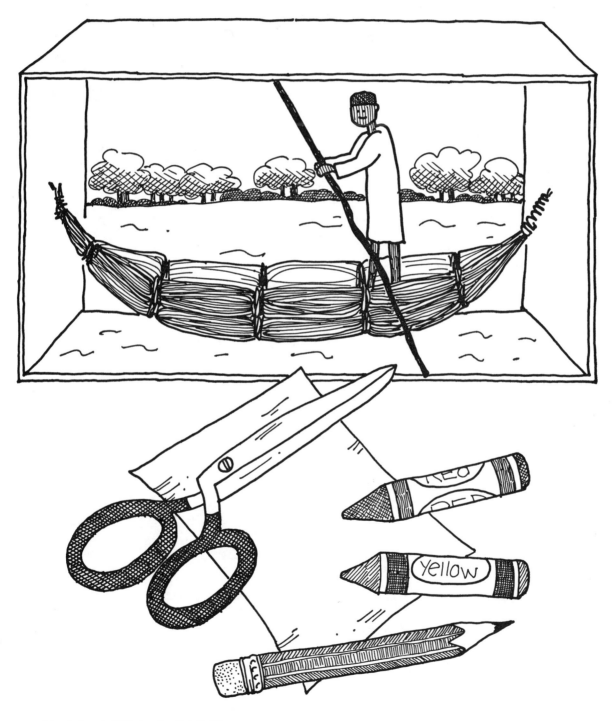 AFRICA

Forms of transportation in Africa differ from region to region because of the diversity of geography. Select a region that interests you. Using a shoebox, make a diorama showing one of the ways the people in that region travel. Print the name of the region on the top of the box and tell why this type of transportation is used.

AFRICA   SOLVE A MYSTERY

For many years, Africa was a mystery to people who lived on other continents. As more and more non-Africans began to visit the large continent, they developed a greater understanding of Africa. Now it's your turn to help solve a mystery. Read a mystery book. Explain how you might have solved the mysterious problem differently.

Name of book: _____

Author: _____

Mystery to be solved:_____

_____

_____

Your solution: _____

_____

_____

_____

_____

_____

Name _____

LITERATURE SECTION | AFRICA

BRINGING THE RAIN TO KAPITI PLAIN
by Verna Aardema

COUNTRY: Kenya
SUMMARY: This is a rhyme explaining how the drought-stricken Kapiti Plain was relieved by rain.

New York: Dial Books, 1981

ACTIVITIES

1. Using the encyclopedia, look up Kenya to find the average yearly rainfall in the plains region.

2. This is a good book for teaching cause and effect. Discuss the meaning of drought. What were its effects on the Kapiti Plain? Illustrate the effects on a poster.

3. Find out more about the acacia tree. Besides providing food for giraffes, how is the tree used?

4. What forms of wildlife live in the plains region? What has the government of Kenya done to protect the animals?

5. Discuss the reasons that animals migrate.

6. Compare this story to the book *THE HOUSE THAT JACK BUILT.* Sequence both stories.

THE VILLAGE OF ROUND AND SQUARE HOUSES
by Ann Grifalconi

COUNTRY: Cameroon
SUMMARY: Grandmother tells her granddaughter about the night the Naka Mountain erupted, leaving only one round house and one square house in the village.

Boston, MA: Little, Brown and Company, 1986.

ACTIVITIES

1. The village of Tos has round houses and square houses. Find out what materials were used to construct the houses. What other types of houses are built in other areas of Cameroon?

2. List the chief crops that farmers in Cameroon grow mainly for their own food. On this list, circle the crops mentioned in the story. What are cash crops? List the cash crops grown in Cameroon. Circle those mentioned in the story.

3. Most volcanic eruptions cannot be predicted. Some, however, seem to have a warning system. Find out about these "warning systems."

4. Use the WORLD ALMANAC to find out where major volcanos have erupted in Africa. Mark the locations on a map. What do these areas have in common?

5. There are three main types of volcanos. Read about each. Use the facts in the story to determine which type erupted in the village of Tos.

6. Construct a miniature volcano.

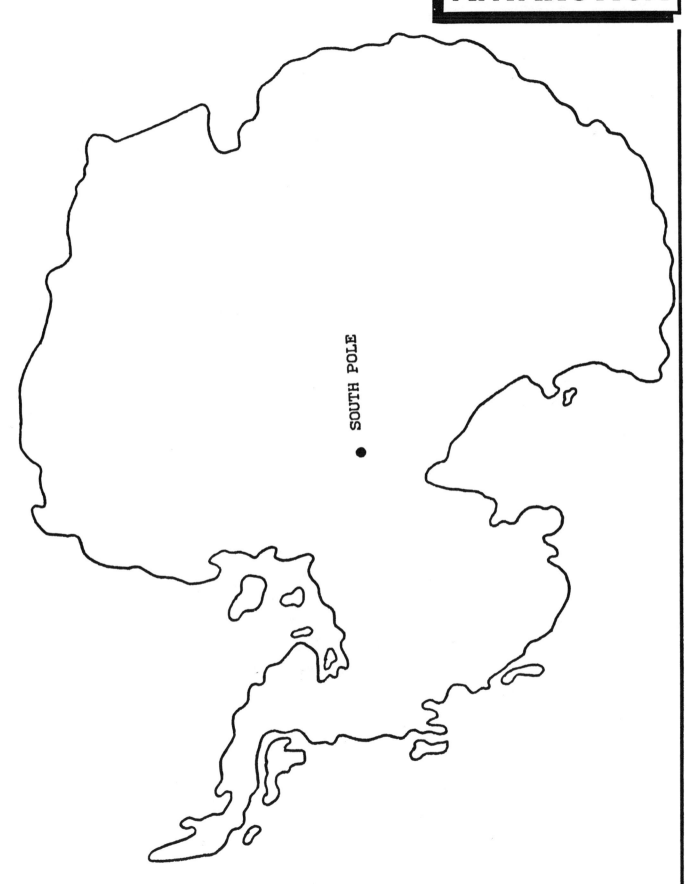

SOUTH POLE

ANTARCTICA OVERVIEW

Antarctica, the coolest place on earth, is the fifth largest continent. This ice-covered area completely encircles the South Pole. The continent is surrounded by the Pacific, Atlantic, and Indian Oceans.

Deposits of metal ores and coal are located on Antarctica, although these sites have not been mined. Petroleum also is believed to be present offshore.

Plant fossils have been found in Antarctica, indicating that a warm climate once existed on the continent. Today, however, only a few plants and insects can survive Antarctica's frozen interior. Penguins, whales, krill, birds, and fish live in the surrounding waters.

People do not live on this continent permanently. Scientific research stations are located on Antarctica. All supplies and food must be delivered during the summer months. The summer season is between November and January. Summer temperatures reach 50 degrees Fahrenheit on the northern islands. During the winter season, May to June, the continent is dark with extremely cold weather and dangerous blizzards.

In 1959, the Antarctica Treaty was signed by twelve countries. It states that Antarctica is to be used for scientific research and exploration only. The findings are to be shared by all participating countries. The research includes such topics as the ocean environment, earthquakes, magnetism, the weather, and the ozone layer above the continent. The treaty forbids military bases, nuclear weapons, and the disposal of radioactive materials. The treaty also protects the plants and animals of Antarctica. Seven of the countries, Australia, France, Argentina, Chile, Great Britain, New Zealand, and Norway, have claimed areas of the continent. These claims, however, are not recognized by the treaty. This peaceful international treaty, for which the United States is the repository, was renewed in 1991. A provision was adopted to ban mining and oil exploration in Antarctica for the next 50 years.

CLAIMS OF ANTARCTICA TERRITORY

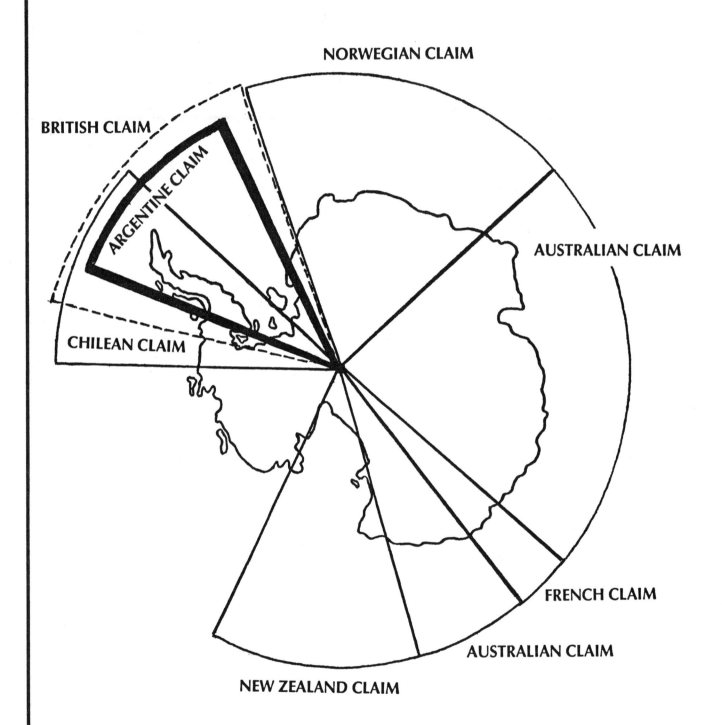

NORWEGIAN CLAIM

BRITISH CLAIM

ARGENTINE CLAIM

CHILEAN CLAIM

AUSTRALIAN CLAIM

FRENCH CLAIM

AUSTRALIAN CLAIM

NEW ZEALAND CLAIM

OUTLINE OF ACTIVITIES | ANTARCTICA

1. **SPEAK OUT!** *(p. 48)*—In 1959, a treaty providing for the protection of Antarctica was signed by twelve nations. Make a list of the provisions of the treaty. State why you feel it is important to maintain specific terms of the treaty.

2. **WHO SAW IT FIRST?** *(p. 49)*—Using a bent coathanger and nylon stocking, make a face of an explorer of Antarctica. Write a paragraph about the explorer's expedition. Select an explorer from the following list:

Roald Amundsen	Lincoln Ellsworth	James Cook
Richard Byrd	Robert F. Scott	Charles Wilkes

3. **CHILLY CREATURES** *(pp. 50-51)*—Select from the list below an animal that lives on Antarctica or in its surrounding waters.

blue whale	humpback whale
crabeater seal	albatross
Emperor penguin	krill

   Assemble a mobile that tells about the animal you have chosen. Include the following facts on the mobile:

a. Picture of the animal	c. Diet	e. Interesting fact
b. Habitat	d. Size	

4. **SURVIVAL NEEDS** *(p. 52)*—You are an explorer being sent to Antarctica. List, in the order of their importance, five items that you ***must*** take for survival in the region. Tell why you would take each.

5. **EXTRA! EXTRA!** *(p. 53)*—Your exploration team has just discovered an interesting fact about Antarctica. Write a news story telling about it. Don't forget the 5W's (who, what, when, where, and why).

6. **A CLAIM TO FAME** *(p. 54)*—Read a biography about a famous person. Design a medallion that represents an accomplishment of this person.

MORE TOPICS TO RESEARCH

Polar jet	Ross Ice Shelf	Enderby Land
IGY—International	Ice cap	Exploration (Polar)
Geophysical Year	Lichens	Antarctic Circle
Krill	Midge	International Whaling
Mt. Erebus	Antarctic Claims	Commission
Ozone	McMurdo Naval	Blizzards
Icebreakers (ships)	Air Station	Antarctic Ocean
Antarctic icebergs	Victoria Land	Southern Elephant
Magnetic South Pole	Transantarctic	Seal
Glaciers	Mountains	Antarctic Peninsula

ANTARCTICA | SPEAK OUT!!!

In 1959, the Antarctic Treaty was signed by twelve countries. It states that Antarctica should be used for scientific research and exploration only. It also protects the animals and plants. This treaty was renewed in 1991. Find out more about the treaty and make a list of its provisions. List five reasons that you feel it is important to maintain (or not to maintain) specific terms of the treaty. Should other provisions be added? Share your opinions with the class.

PROVISIONS OF ANTARCTIC TREATY

_____
_____
_____
_____
_____
_____
_____
_____

REASONS TO UPHOLD (OR TO NOT UPHOLD) TERMS OF TREATY

1. _____
2. _____
3. _____
4. _____
5. _____

Name _____

WHO SAW IT FIRST? | ANTARCTICA

Before the continent of Antarctica was discovered, ancient Greek philosophers believed that a large mass of land was needed on the southern end of the Earth to balance the other land regions. It was not until the late 1700's that the search for the continent began. Although the records are not clear as to who first saw the continent, many explorers have led important expeditions and made important discoveries.

Select an explorer from the list below and write a paragraph to describe his expedition.

Cover a bent coathanger with a nylon stocking and make a face of the explorer. Attach your report.

WHO SAW IT FIRST?

Roald Amundsen	Robert F. Scott
Richard Byrd	James Cook
Lincoln Ellsworth	Charles Wilkes
James C. Ross	Nathaniel B. Palmer

ANTARCTICA | CHILLY CREATURES

Because of the cold climate and lack of plant life, few animals live on the continent of Antarctica. Find out about the animals that do live there by assembling an animal mobile.

Select an animal from the list below. Complete the information cards, cut them out, and tie them onto a hanger. Decorate the hanger to resemble a chunk of ice.

ANTARCTIC WILDLIFE

albatross   crabeater seal   humpback whale

blue whale   Emperor penguin   krill

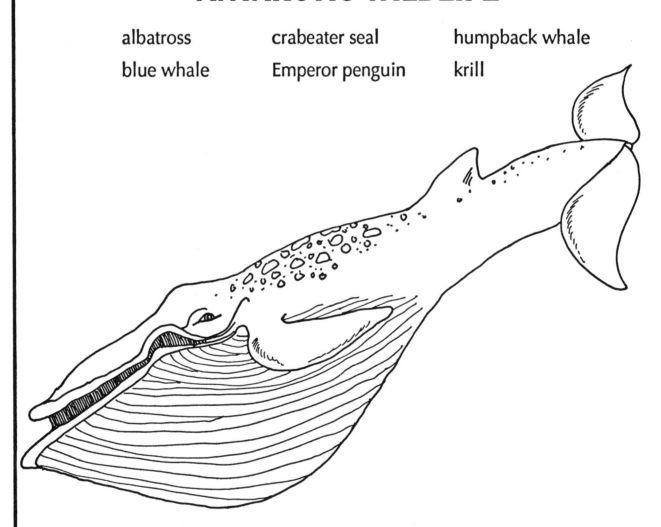

CHILLY CREATURES ANTARCTICA

ANTARCTICA SURVIVAL

Before expeditions to Antarctica, there was no information about the continent. Some people believed the land was populated and had plant life. Others thought that the heat around the equator would keep explorers from traveling south. The first explorers, however, were blocked by huge pieces of ice and didn't see land.

Let's pretend that you are a member of one of the teams sent in the 1800's to explore Antarctica. List, in order of importance, five items that you MUST take for survival in this region. Tell why you would need each item. Remember—there is no electricity!

1. _____

   Why: _____

2. _____

   Why: _____

3. _____

   Why: _____

4. _____

   Why: _____

5. _____

   Why: _____

EXTRA!! EXTRA!! READ ALL ABOUT IT!!

ANTARCTICA

The continent of Antarctica is used mostly for scientific research. You are a member of the United States team located at the McMurdo Station on Ross Island. The researchers recently discovered an interesting fact about Antarctica.

You have been assigned to write a news story telling about the discovery. Complete the information below before writing the article.

THE FIVE W'S

1. Who:_____

_____

2. What: _____

_____

3. When:_____

_____

4. Where: _____

_____

5. Why:_____

_____

A CLAIM TO FAME

Throughout history, there have been many great people who have contributed to making the world a better place. Read a biography about a famous person. Design a medallion to show recognition of one of this person's major accomplishments.

Name of book: _____

Author: _____

Famous person: _____

Major accomplishment: _____

ANTARCTICA
by Helen Cowcher

CONTINENT: Antarctica

SUMMARY: The creatures of Antarctica share a strange, beautiful, and quiet continent. Their main enemies are each other. Survival requires constant watching for larger creatures who are stalking for food . . . until new creatures, humans, arrive to share their world.

New York: Farrar, Straus and Giroux, 1990

ACTIVITIES

1. Explain why it is dark both day and night during the winter months on the continent of Antarctica. During which months does winter occur?

2. Research the eating habits of each of the life forms found in and around Antarctica. Make a list beginning with the smallest, algae, and concluding with the largest, the blue whale. Construct a chart to illustrate the Antarctic food chain.

3. Compare Adelie penguins to Emperor penguins.

4. Find out more about the importance of icebreakers in the Antarctic region.

5. Brainstorm ways the new creatures, humans, could destroy or share Antarctica with its wildlife.

A TALE OF ANTARCTICA
by Ulco Glimmerveen

CONTINENT: Antarctica
SUMMARY: Through the eyes of penguins, the reader sees how pollution is destroying the continent of Antarctica.

New York: Scholastic Inc., 1990

ACTIVITIES

1. List the items that litter the beaches as described in the book. What types of pollution are destroying beaches throughout the world?

2. Simulate an oil spill by pouring a small amount of oil into a shallow pan of water. Drop shells, rocks, sand, feathers, and other items into the water. Think of ways to clean these items. How can the "spill" be removed? After discussing different methods, try removing the oil with spoons to scoop it up, paper towels to absorb it, and wind to blow it away. Which method works best? List the pros and cons of each method.

3. Find out why penguins live only in the southern part of the world.

4. The Emperor penguin has unusual nesting habits. What are the responsibilities of the male and female after the female lays an egg?

5. After brainstorming about the environmental problems of the world, write a paragraph describing the perfect environment in which man could live.

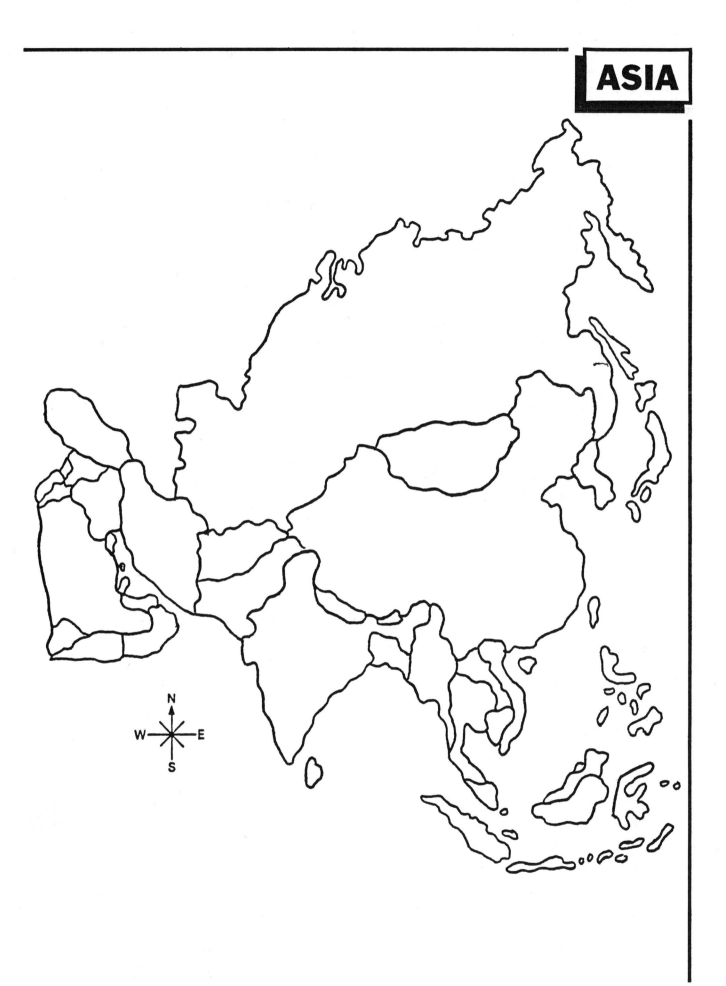

ASIA OVERVIEW

Asia is the largest continent in both size and population. It extends from Africa and Europe on the east to the Pacific Ocean on its west. The northern regions are in the Arctic, while the equator passes through the southern, tropical regions. Both the highest and lowest points on Earth are located on this continent: Mount Everest is the highest point and the Dead Sea is the lowest.

Civilization had its beginnings in Asia. All of the world's major religions also began on this continent. Nine language families are spoken in this part of the world, and agriculture is the most important economic activity.

Southwest Asia, also known as the Middle East, is mainly a desert region. The people live crowded along coastlines and waterways. The area lacks good farmland; however, half of the people are farmers. Since water is scarce, the people must irrigate the land on which barley, wheat, dates, and oranges are grown. The area is rich in oil. The Muslim religion, Islam, strongly influences the art of the region. Mosques, rugs, textiles, and ceramics all reflect the detailed patterns of Islamic art. Islam forbids the use of animals or human impressions on works of art.

South Asia is the world's most crowded area. This poverty-stricken region has much fertile land and is beset by monsoons. Tea is the most important crop. Barley, millet, rice, and beans are also grown. The Hindu religion is very important to the lifestyle of this area. Hinduism prohibits the eating of meat. The people are divided into social classes by a caste system. The Taj Mahal, a

magnificent example of Islamic architecture, and Mt. Everest, the highest point in the world, are located in this region.

Southeast Asia has fertile soil and valuable minerals. Much of the land is mountainous. Crops of coffee, rubber, sugar cane, tea, and tobacco, as well as vegetables and fruits, are grown here. Fish are plentiful in the surrounding waters. Floating markets are common on the waterways. Buddhism is the main religion of the mainland. A religious temple called a pagoda can be found in most villages.

East Asia is densely populated. Because of the large population, many people live on houseboats. Tibet, the highest plateau in the world, can be found in the mountain region of China. The people grow rice on the rich farmland. The Chinese invented the compass, paper, and printing. The other countries of this region are industrial nations. Japan is a modern country with many western ways. It is the world's largest producer of radios, cameras, cars, and televisions. Baseball is the favorite sport. The beauty of the land is reflected in Mt. Fuji.

North Asia lies mostly in Siberia, an area of long, bitter winters. Because of the cold climate, large masses of land are underdeveloped. Much of the land is covered by forests. Some of the people herd sheep, cattle, and reindeer, while others grow crops. The region is a leading wheat-producing area.

Central Asia consists of grassy plains, huge deserts, and mountains. In this sparsely-populated area, most people herd yaks and sheep. The land is too dry and rugged for farming.

INDEPENDENT

Afghanistan
Bahrain
Bangladesh
Bhutan
Brunei
Burma
China
* Commonwealth of
Independent States (Asian):
    Kazakhstan
    Kyrgyzstan
    Russia
    Tajikistan
    Turkmenistan
    Uzbekistan
Cyprus
India
Indonesia (Asian)
Iran
Iraq
Israel
Japan
Jordan
Kampuchea
Kuwait
Laos
Lebanon

Malaysia
Maldives
Mongolia
Nepal
North Korea
Oman
Pakistan
Philippines
Qatar
Saudi Arabia
Singapore
South Korea
Sri Lanka
Syria
Taiwan
Thailand
Turkey (Asian)
United Arab Emirates
Vietnam
Yemen (Aden)
Yemen (Sana)

DEPENDENT

British Indian Ocean Territory
Gaza Strip
Hong Kong
Macao
West Bank

* Each member of the Commonwealth is recognized as an independent nation.

OUTLINE OF ACTIVITIES ASIA

1. **MUSIC: THE UNIVERSAL LANGUAGE** *(p. 62)*—Asian music is very different from the music of western cultures. Draw a picture of the Asian instruments listed below and write a paragraph describing each.

   koto   shakuhachi   jew's harp   tambourine
   gong   sitar   samisen

   Select two of the instruments. List three ways these are alike and different.

2. **LEARN ABOUT THE ECONOMY** *(p. 63)*—Select an Asian country. On the sides of a cube, illustrate aspects of the economy of the country. These should include either the agriculture, mining, manufacturing, fishing, exports, or forestry of the country.

3. **ASIAN CROSSWORD PUZZLE** *(pp. 64-65)*—Using the World Almanac, find the facts about Asia to complete the crossword puzzle. (The answers are found on page 65.)

4. **GO FLY A KITE!** *(p. 66)*—Kites originated in China over 3000 years ago. Read about kites, and learn more about their appearances and other uses. Next, write a story about a boy and how the kite helped him perform a good deed. Illustrate your story.

5. **ASIAN LANDMARKS** *(p. 67)*—Many people visit Asian countries to see specific landmarks. Select one of the landmarks listed below and write a paragraph describing it. Tell where and when it was built. Include other interesting facts. Draw a picture of the landmark.

   Great Wall of China   Shwe Dagon Pagoda   Mt. Fuji
   The Great Mosque in Mecca   Wailing Wall   Taj Mahal

6. **POETRY: IMAGES OF NATURE** *(p. 68)*—List three ways that a poetry book is different from a fiction book. There are two types of Japanese poetry: haiku and tanka. Write a poem that is an example of one of these forms.

MORE TOPICS TO RESEARCH

Trans-Siberian Railroad	Dyaks	Mecca
	Komodo Dragon	Chinese New Year
Tarsier	Mt. Fuji	Origami
Panda	Great Wall of China	Rice production
Islamic Art	Junk (boat)	Judo
Caste System	Monsoon	Chinese Ideographs
Jute	Pagoda	Japanese Chin
Bamboo	Silk production	Calligraphy
Rajah	Tea plant	

MUSIC—The Universal Language

Asian music is very different from the music of western cultures. Not only are the instruments different, but the scales used to compose the music are not the same. Much Asian music has no harmony.

Find out more about the instruments used in Asian countries. Fold a large piece of paper (12" x 18") into eight sections. Using only the top half, draw a picture of each instrument listed below. Continue your drawings on the back side. Below each picture, write a paragraph to describe the instrument.

ASIAN INSTRUMENTS

| koto | sitar | shakuhachi | tambourine |
| gong | samisen | jew's harp | |

Select two of the instruments. List three ways they are alike and different.

Instruments: _____ & _____

	ALIKE	DIFFERENT
1.	_____	_____
2.	_____	_____
3.	_____	_____

Name _____

LEARN ABOUT THE ECONOMY   ASIA

The economy is the way a country produces, consumes, and distributes its resources. Every country has an economy. Some are highly developed and provide great wealth to the country, while others are undeveloped and poor.

Select a country on the continent of Asia. After covering the sides of a box with paper, illustrate four different aspects of that country's economy. These should include either the agriculture, livestock, mining, manufacturing, fishing, exports, or forestry activities.

Directions: Use the WORLD ALMANAC to find the answers to the clues for the puzzle.

ACROSS CLUES
4. A major earthquake shook _____, China on December 16, 1920.
5. A major industry and mineral found in Saudi Arabia is _____.
7. The capital of Israel is _____.
9. A typhoon hit Japan and South Korea on September 17-19, 1959. Its name was _____.
10. The currency used in Turkey is called the _____.

DOWN CLUES
1. The leading exporter of rice is _____.
2. The city in Indonesia with the largest population is _____.
3. The official language of Burma is _____.
6. The world's longest railway tunnel, located in Japan, is the _____ Tunnel.
8. The deepest lake in Asia is the _____ lake.

Name _____

CROSSWORD PUZZLE · ASIA

ANSWERS
to Crossword Puzzle on Page 64.

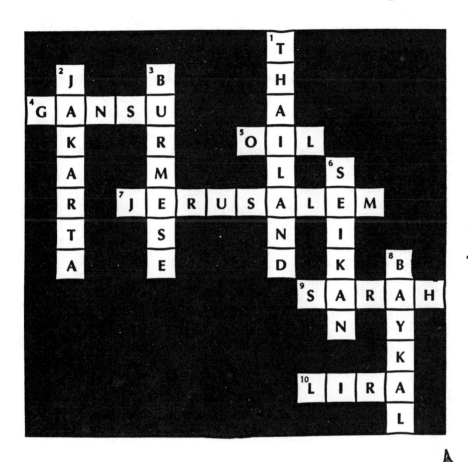

WORD LIST

BAYKAL	LIRA
BURMESE	OIL
GANSU	SARAH
JAKARTA	SEIKAN
JERUSALEM	THAILAND

GO FLY A KITE!

Kites, named after the graceful kite bird, originated in China over 3000 years ago. Brightly painted kites were used to signal soldiers, scare birds away from crops, catch fish, and frighten bandits from houses. Today people enjoy flying many types of kites as a recreational activity.

Read about kites and learn more about their appearances and uses. Describe what you learn on the lines below. Next, on another sheet of paper, write a story about a boy and the kite that helped him perform a good deed. Illustrate your story.

Types of kites:

1. _____
2. _____
3. _____
4. _____
5. _____

Other ways kites have been used:

1. _____
2. _____
3. _____
4. _____
5. _____

Name _____

ASIAN LANDMARKS  ASIA

The landmarks of Asia represent the natural beauty of the land, historical contributions, architectural marvels, and the cultures of the people.

Choose an Asian landmark and write a paragraph to describe it. State where and when it was built, and include other interesting facts. Draw the famous site on a separate piece of paper.

FAMOUS ASIAN SITES

Great Wall of China   Mt. Fuji

Wailing Wall   Shwe Dagon Pagoda

The Great Mosque in Mecca   Taj Mahal

_____

_____

_____

_____

_____

_____

_____

_____

_____

_____

_____

Name _____

The Japanese have given us two forms of poetry that focus on nature: haiku and tanka. Haiku is a nonrhyming poem of three lines, while a tanka has five lines. Each line contains a specific number of syllables. Write a verse of tanka or haiku using colorful and expressive phrases.

TANKA

Line 1 = 5 syllables

Line 2 = 7 syllables

Line 3 = 5 syllables

Line 4 = 7 syllables

Line 5 = 7 syllables

HAIKU

Line 1 = 5 syllables

Line 2 = 7 syllables

Line 3 = 5 syllables

Name _____

LITERATURE SECTION ASIA

THE MAGIC FAN
by Keith Baker

COUNTRY: Japan
SUMMARY: A young boy uses his talents to build a bridge that saves the people of his village from a tidal wave.

New York: Harcourt Brace Jovanovich, 1989

ACTIVITIES

1. Find out about tidal waves, or the tsunami, as the Japanese call it. How are they different from hurricanes?

2. Find examples of oriental fans and kites. How are they alike and different?

3. Construct a kite. Have a Japanese kite festival. Make this a goal-setting day. Just as the young boy built items to reach his goals, write a personal goal and attach it to your kite.

4. On a fan-shaped piece of paper, write a haiku poem. Focus on a subject of nature from the story (sea, moon, rainbow, clouds, sky, wind, tsunami, etc.).

5. The village in the pictures of the story is very crowded. There appears to be very little space between the houses. Look at a map of Japan to determine its size. Find the population of Japan. Japan is the world's seventh largest nation in population. Brainstorm ways that future generations of Japan can use the limited space to maintain or improve the present living conditions.

6. Compare this story to *THE HOLE IN THE DIKE* by Norma Green.

THE BIRD WHO WAS AN ELEPHANT
by Aleph Kamal

COUNTRY: India
SUMMARY: A bird, who had been an elephant in another life, revisits a small village in India and observes the lifestyles of the people.

New York: J.B. Lippincott, 1989.

ACTIVITIES

1. Make a picture dictionary of the Indian words used in the story.

2. Explain, according to Hindu beliefs, how the bird could once have been an elephant.

3. List the Indian occupations mentioned in the book. Write a short paragraph telling about one of them.

4. Choose one of the spices that is sold in the spice shop. Find out how it is grown, processed, and used.

5. The palmist told the bird that he had once been an elephant that carried children across the palace gardens of the Maharajah. What is a Maharajah?

6. On special occasions the palace elephants were decorated with jewels and tapestries to represent the wealth of the Maharajah. Draw an elephant and carefully decorate it. Pretend you are a child of the Maharajah. Write a story about your life in the palace.

7. Elephants help the environment. List the ways they help. Which of these do you consider to be the most important? Why?

AUSTRALIA

AUSTRALIA OVERVIEW

Australia is the smallest continent and the only continent that is also a country. Australia lies entirely in the Southern Hemisphere and is located between the Indian and Pacific Oceans, and southeast of Asia. Because of its location under the equator, the continent is sometimes referred to as being "down under." Australia is divided into six states and two territories.

Australia is the world's driest continent. It is covered mainly by deserts and dry grasslands. Artesian wells are an important source of water. Most of the people live along the eastern coastline and are of British ancestry.

Australia has three main land areas. The Eastern Highlands area, sometimes called the Great Dividing Range, is located along the eastern coastline. It extends into Tasmania and is the most populated area of the continent. This region is covered by rich farmland. The Australian Alps, a mountain range formed by volcanic eruptions, also extend through this region.

The Central Lowlands area is generally a flat region. Wheat is grown in the southern part of this region. The rest of the region is too hot and dry to grow crops. The dry, grassy area is good for grazing. Desert land also exists in this region.

The Western Plateau covers two thirds of Australia. Deserts cover the central area, while grassy, grazing areas are found in the northeastern part. One third of Australia is covered by deserts. One of the most spectacular land formations, the Ayers Rock, is located in this region. Paintings by the Aborigines, the first known inhabitants of Australia, cover the walls of caves in this formation.

The Great Barrier Reef, the world's largest coral reef, is located off the northeast coastline. Over 400 species of coral and hundreds of rare forms of sea life inhabit this reef, which stretches over 1200 miles.

Many unique animals live on this continent. They include the echidna and platypus, which are egg-laying mammals. Kangaroos, koalas, wombats, and wallabies are marsupials that live in Australia.

Australia is the world's leading exporter of wool. There are nine sheep to every person on this continent. Other major farm products include cattle, wheat, sugar cane, and fruit. The continent also is a leading producer of bauxite. Most of the world's high-quality opals come from Australian mines.

AUSTRALIA REGIONS

STATES

New South Wales
Queensland
South Australia
Tasmania
Victoria
Western Australia

MAINLAND TERRITORIES

Australian Capital Territory
Northern Territory

OTHER TERRITORIES

Ashmore and Cartier Islands
Australian Antarctic Territory
Christmas Islands
Cocos (Keeling) Islands
Coral Sea Islands
Heard and McDonald Islands
Norfolk Island

OUTLINE OF ACTIVITIES | AUSTRALIA

1. **AS DIFFERENT AS NIGHT AND DAY** *(p. 76)*—Complete the chart contrasting the two continents. Remember—contrasting means to show noticeable differences.

2. **IT'S HOW YOU PLAY THE GAME** *(p. 77)*—There are many popular team sports played in Australia. Choose one of the sports and write a paragraph to describe the game to a friend. Include the number of players required and the rules.

           Cricket   Soccer   Rugby

3. **"EIGHTH WONDER OF THE WORLD"** *(p. 78)*—The Great Barrier Reef is sometimes called the 8th wonder of the world. Read about this special place. Choose one life form that makes the reef its home. Make a cutout of the species and attach it, along with a short descriptive paragraph, to a class mural.

4. **S.O.S.** *(p. 79)*—You have been shipwrecked along the coast of Australia in the town of _____ . Write an SOS message. Include your location (town, latitude, and longitude), the crops that are grown nearby, and at least two types of native animals that you have seen.

5. **SAY "CHEESE"** *(pp. 80-81)*—You have been selected to "photograph" a book for a major magazine. Complete two "photos" of events from the story that definitely would not happen in real life. Complete two "photos" of events that could happen in real life. If you read a realistic fiction book, complete four "photos" of events that could happen in real life.

6. **ABORIGINAL ART** *(p. 82)*—Find examples of Aboriginal art. Paint an "Aboriginal" design of your own. If possible, create your art work on a rock or piece of wood. Include a native animal of Australia in your design.

MORE TOPICS TO RESEARCH

Aborigines	Emu	Colombo Plan
Boomerang	Ayers Rock	Australian Terrier
Sheep raising	Mt. Kosciusko	Bottle tree
Coral Sea	Seasons	Polyp
Marsupials	Australian Cattle Dog	Indian Ocean
Great Victoria Desert	Platypus	Natural pearls
Tasman Sea	Snowy Mountain	Education (in remote
ANZUS	Scheme	areas)
Australian Flag	Royal Flying Doctor	
The language	Service	

AUSTRALIA AS DIFFERENT AS NIGHT AND DAY

List 10 ways these continents are different.

ANTARCTICA

1. _____
2. _____
3. _____
4. _____
5. _____
6. _____
7. _____
8. _____
9. _____
10. _____

AUSTRALIA

1. _____
2. _____
3. _____
4. _____
5. _____
6. _____
7. _____
8. _____
9. _____
10. _____

Name _____

IT'S HOW YOU PLAY THE GAME

AUSTRALIA

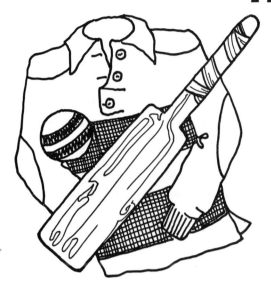

The climate of Australia is very mild in most areas. This makes it possible for the people to enjoy many types of sports. Australians are highly competitive and believe strongly in good sportsmanship. Since most of the people live near the coast, Australians enjoy a variety of water sports. Some of the most popular activities, however, are the team sports of cricket, soccer, and rugby. These games were brought to Australia by settlers from Great Britain.

Select a team sport played in Australia. Write a paragraph describing the game. Include the number of players required and the rules of the game. Share your description with a friend.

_____

_____

_____

_____

_____

_____

_____

_____

_____

_____

_____

Name _____

AUSTRALIA 8TH WONDER OF THE WORLD

The Great Barrier Reef, sometimes called the "eighth wonder of the world," is the largest chain of coral reefs on earth. Some scientists believe that the reef is millions of years old. It is located off the northeast coast of Australia. A large variety of sea life inhabits the reef. It is illegal to take any of the coral.

Read about this special place. Choose one life form that makes the reef its home. Make a cutout of the species and glue it to a piece of paper on which you have written a short descriptive paragraph. Attach your paper to a wall or bulletin board to form a class mural.

All oceans have currents. These are the general movements of the water. To find out more about currents and the directions they travel, scientists throw drift bottles into the oceans and trace their routes. Bottles have also been used to signal for help. The notes in the bottles are called S.O.S. (Save Our Ship) messages.

You have been shipwrecked along the coast of Australia in the town of _____. Write an S.O.S. message on the "bottle." Include your location (town, latitude, and longitude), the crops that are grown in the region, and two native animals that you have seen.

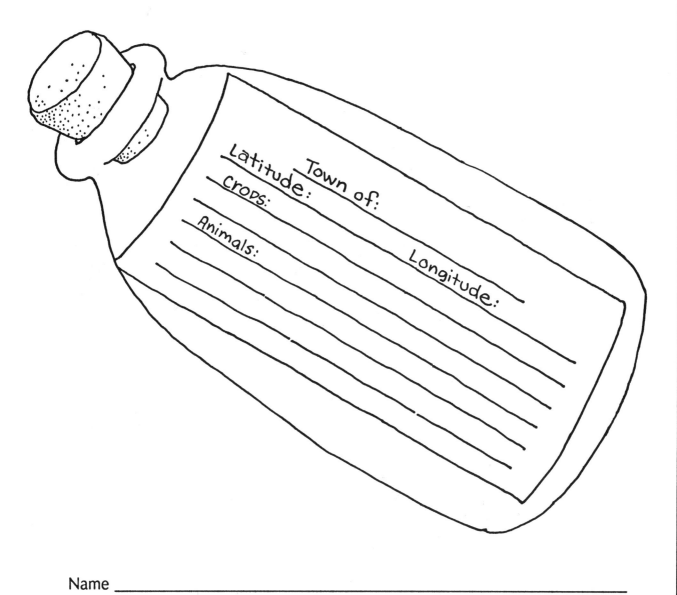

Town of:

Latitude:

Crops:

Animals:

Longitude:

Name _____

SAY CHEESE

You have been selected to "photograph" a fiction book for a major magazine. After reading the book, complete two "photos" of events from the story that would definitely not happen in real life. Complete two "photos" of events that could happen in real life. If you read a realistic fiction book, complete four "photos" of events that could happen in real life. Write a caption for each picture.

COULD HAPPEN IN REAL LIFE

Name _____

COULDN'T HAPPEN IN REAL LIFE

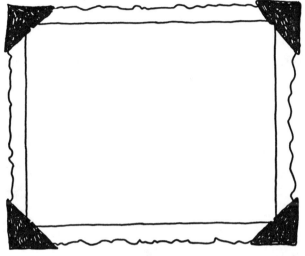

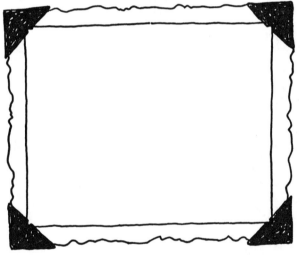

Name of book: _____

Author of the book: _____

Main characters: _____

Setting: _____

Name _____

AUSTRALIA | ABORIGINAL ART

The Aborigines were the first people to live in Australia. Since these early inhabitants had no written language, they depicted rituals, beliefs, and animals through their primitive art. Most of the paintings were completed on eucalyptus bark, stones, and cave walls. The drawings were simple and composed of geometric shapes. The Aborigines used the designs to decorate boomerangs, utensils, shields, and weapons.

Find examples of these designs. Then paint an Aboriginal drawing of your own. Create your art work on a rock or piece of wood. Include a native animal of Australia in your design.

LITERATURE SECTION | AUSTRALIA

WHERE THE FOREST MEETS THE SEA
by Jeannie Baker

CONTINENT: Australia
SUMMARY: On a boat trip with his father to a remote area of Australia, a young boy recalls the history of the Daintree Rain Forest and imagines what the forest will be like in the future.

New York: Greenwillow Books, 1987.

ACTIVITIES

1. In several of the pictures in this book, there are faint images of life forms that once inhabited the forest. Choose one of the life forms and write a story describing its feelings about the changes it saw taking place in the forest. How did the changes effect its way of life?

2. As the boy begins to walk through the forest, the author tells us that "now the forest is easy to walk in." As a class, brainstorm the following questions:
   a. What does the author's statement mean?
   b. What do you believe a walk through the forest was like in the past?
   c. How has it changed?

3. The young boy stops to wonder how long it would take for a tree in the forest to become fully grown. Why do you think the author draws our attention to this thought? Find out how trees grow from seeds. Draw a diagram to illustrate the process. How can the age of a tree be determined?

4. The first people to live in Australia were the Aborigines. Compare their traditional culture to their present way of life.

5. As the trip with his father ends, the boy imagines the future of the forest. What does he envision it to be like? Make a list of the pros and cons of his vision.

POSSUM MAGIC
by Mem Fox

CONTINENT: Australia
SUMMARY: Invisible Hush and Grandma Poss travel around Australia eating "people" food in hopes of making Hush visible again.

New York: Gulliver Books, 1990.

ACTIVITIES

1. Hush and Grandma Poss visited seven capital cities trying to find the right food to make Hush visible. Name the area of which each city is the capital. Six of these areas are states of Australia. Which area is not? Why is it not a state?

2. On a map of Australia, mark each capital city and trace the trip the two possums took. Name at least one place of interest in each state that you would like to visit and tell why.

3. Other than the foods mentioned in the story, what do the people of Australia like to eat?

4. As a math lesson, make lamington (a square of sponge cake dipped in chocolate and covered in coconut). Use a rectangular pan to bake the cake. Determine how to cut the cake so that each piece is equal in size and everyone in the class receives a piece.

5. Several animals that are native to Australia are mentioned in the story. Divide the class into groups. Instruct each group to display the natural habitat of one of these animals by standing a large cardboard box on its side and decorating the interior. Place a large cutout of the animal and general information about it in the box. Which of the animals are marsupials?

EUROPE

EUROPE OVERVIEW

Europe, a highly industrialized world region, is the second-smallest continent. As the birthplace of western civilization, Europe holds a prominent place in world history and present-day affairs. It is the most densely populated continent. Approximately 50 languages are spoken here by more than 150 ethnic groups. Sightseers from all areas of the world travel to this continent to visit its historical landmarks and enjoy the natural beauty of the land.

Northern Europe, or the Norden region, is the coolest part of the continent. The winters are long and the summers are short. This region includes the Scandinavian countries of Norway and Sweden, as well as Finland, Denmark, Iceland, and the northern regions of Russia. The land consists of plateaus, highlands, and the rugged Northwestern Mountains. It is the most sparsely populated region of Europe. Although the land is good for grazing, most of it is unproductive. Fishing, shipbuilding, forestry, and dairy farming are the main occupations.

Western Europe is a heavily populated, mostly industrial region. It is a center for world trade because of its location on major waterways. The region has abundant iron ore and coal. The landscape of the area has fertile plains and picturesque mountains. The world's most fertile farmland lies in the North European Plain. Crops of grapes, potatoes, barley, and wheat are grown. This rich land extends from southeast

England, the Netherlands, Belgium, Luxembourg, through northwest France, Germany, and into the western European countries of Poland, Belarus, and Russia. The polders of the Netherlands, the Black Forest of Germany, and the Swiss Alps are located in this region. Tourism is also a major source of income.

Eastern Europe is sandwiched between the Baltic Sea on the north and the Adriatic and Black Seas on the south. The agricultural crops of this region include rye, potatoes, wheat, and corn. Livestock is also raised.

The land of most southern European countries is hilly and mountainous with poor soil. Spain and Portugal are located on the Iberian peninsula on a large, high, and dry plateau. The people grow crops of grapes and olives. Bullfights are a popular attraction. Italy, a boot-shaped country, Vatican City, and San Marino are the countries that occupy the Italian peninsula. Grapes are Italy's most valuable crop. Shoes rank as a leading manufactured product. Italy was the center of the ancient Roman Empire. The country of Greece, home of the ancient Greek civilization, is found on the Balkan peninsula. Cotton and tobacco crops are grown in this region. Greece is a leading producer of lemons, raisins, and olives. Although this is one of the poorest parts of Europe, the coastlines of this country are popular with tourists.

EUROPE COUNTRIES

INDEPENDENT

Albania
Andorra
Austria
Belgium
Bosnia-Herzegovina
Bulgaria
* Commonwealth of
  Independent States (European):
  Armenia
  Azerbaijan
  Belarus
  Moldova
  Russia
  Ukraine
Croatia
Czech and Slovak Federal Republic
Denmark
Estonia
Finland
France
Georgia
Germany
Great Britain
Greece
Hungary
Iceland
Ireland

Italy
Latvia
Liechtenstein
Lithuania
Luxembourg
Malta
Monaco
Netherlands
Norway
Poland
Portugal
Romania
San Marino
Slovenia
Spain
Sweden
Switzerland
Turkey (European)
Vatican City
** Yugoslavia

DEPENDENT

Azores
Channel Islands
Faeroe Islands
Gibraltar
Isle of Man

* Each member of the Commonwealth is recognized as an independent nation.
** The Republic of Macedonia considers itself to be an independent nation from Yugoslavia; however, its independence has been recognized officially only by the nation of Bulgaria.

OUTLINE OF ACTIVITIES | EUROPE

1. **LIGHTS, CAMERA, ACTION!** *(p. 90)*—Select a European country. Locate ten facts about life in that country. Using the researched information, create a "filmstrip." Don't forget the title frame, credit frame, and THE END.

2. **COMMEMORATIVE STAMP** *(p. 91)*—Each European country has important dates, festivals, holidays, and events to remember. Select a country, read about it, and design a stamp to commemorate a special event. The price on the stamp must reflect the currency of the country.

3. **ALL ABOARD!** *(p. 92)*—Select five landmarks that you would like to visit from the list below. (You may choose only one from each country.) Find a map of Europe. Measure the distance in miles and kilometers between the landmarks.

   France....................Arc de Triomphe OR Eiffel Tower
   Great Britain ....Buckingham Palace OR Big Ben
   Greece...................Parthenon
   Italy ....Gondolas in Venice OR The Colosseum OR
   Leaning Tower of Pisa
   Netherlands ....The Polders in Amsterdam
   Spain ....Bullfights in Madrid
   Switzerland............Swiss Alps in St. Moritz
   Russia ....St. Basil's Church OR Kremlin

4. **SOUVENIR CARDS** *(pp. 93-95)*—While visiting the different countries of Europe, you have purchased six souvenirs to bring home to your family and friends. Each gift represents the lifestyle or craftsmanship of a country. Learn more about the items and the people of the country represented by each as you complete the set of souvenir cards.

5. **ONCE UPON A TIME** *(p. 96)*—Many of our familiar folktales came from Europe. Think of another ending for the story.

MORE TOPICS TO RESEARCH

The Black Forest	Gondola	Danube River
Polders	Loch Ness Monster	Flamenco
Marble	Lapland	Cork Oak tree
Opera	Arc de Triomphe	Tulips
Cheesemaking	Bastille Day	Blarney Stone
Matterhorn	Berlin Wall	Simplon Tunnel
Leaning Tower of Pisa	Olives	The ancient
Windsor Castle	Pasta	Olympic Games
Dikes	Parthenon	

Although about forty-five countries occupy the European continent, each one is very different from the others. Not only are the land regions different, but so are the people, plant and animal life, and histories.

As a movie director, visit a European country. Make a list of ten facts that would interest others. Divide a 4-inch wide strip of paper into fifteen 5-inch sections. Create a "filmstrip" using the facts, and write a script. Don't forget the title frame, credit frame, and THE END. Leave blank frames at the beginning and end. Cut a 4-inch by 5-inch opening in the bottom of a shoebox. Stand the box on its end and use it for your "filmstrip" viewer.

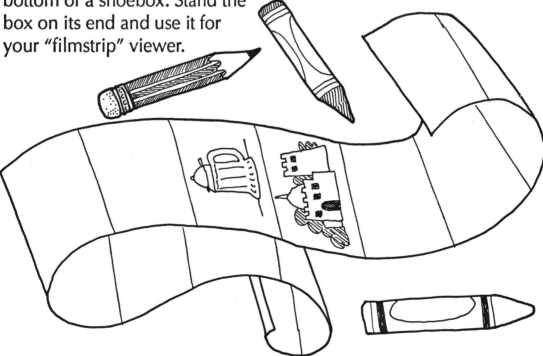

COMMEMORATIVE STAMP | EUROPE

Each European country has important dates, festivals, holidays, and events to remember. Select a country. After reading about it, design a stamp to commemorate a special event, person, or era in the history of that country. Don't forget to put the price of the stamp on your design. The price should reflect the currency of that country.

Name _____

ALL ABOARD!

Tourists flock to Europe each year to visit its famous sites. From viewing the changing of the guards at Buckingham Palace to skiing the Swiss Alps, there are sights and activities for everyone to enjoy.

Tour Europe on the "European Express." From the list below, select five landmarks that you would like to visit. Find a map of Europe and plan a route from landmark to landmark. Measure the distances between the landmarks in both miles *and* kilometers. Record your findings below. The "European Express" departs from Billund, Denmark.

France .... Arc de Triomphe *or* Eiffel Tower
Great Britain ... Buckingham Palace *or* Big Ben
Greece .... Parthenon
Italy .... Gondolas in Venice *or* The Colosseum
                        or Leaning Tower of Pisa
Netherlands ..... The Polders in Amsterdam
Spain............... Bullfights in Madrid
Switzerland...... Swiss Alps in St. Moritz
Russia .... St. Basil's Church *or* the Kremlin

From: Billund, Denmark

Landmark	Miles	Kilometers

To: _____

To: _____

To: _____

To: _____

To: _____

Name _____

SOUVENIR CARDS | EUROPE

While visiting the different countries of Europe, you have purchased seven souvenirs to bring home to your family and friends. Each gift represents the lifestyle or craftsmanship of a country. Learn more about the items and the people of the country represented by each as you complete the set of souvenir cards.

SWISS WATCH
from
SWITZERLAND

Switzerland is well known for its watchmaking industry. The skilled craftsmen pride themselves on their precision and engineering accomplishments. Over 95% of the watches made in Switzerland are exported to other countries. However, most of the raw materials used to manufacture such items as watches must be imported from other countries. Name two reasons why Switzerland must import most of the resources needed for manufacturing.

1. _____
   _____
2. _____
   _____

Name_____

A KILT
from
SCOTLAND

In Scotland, one of the four countries that are part of Great Britain, families with the same name or ancestors are called clans. Each clan has its own plaid pattern which is used on kilts. A kilt is a knee-length pleated skirt. The plaid pattern is called a _____.
Name three other pieces of clothing that are considered to be part of the Scottish traditional dress.

1. _____
2. _____
3. _____

Name_____

SKALLER
from
FINLAND

Skaller are warm, ankle-high moccasins made of reindeer skin and insulated with grass. They are worn by the people who live in the northern or Lapland region of Finland. What other uses do the Lapps have for reindeer?

1. _____
_____

2. _____
_____

3. _____
_____

4. _____
_____

Name_____

CUCKOO CLOCK
from
GERMANY

Many cuckoo clocks are made by the people who live in the Black Forest, a mountain region of Germany. This region, through which the Danube River flows, is covered with forests of dark fir and spruce trees. Toys, musical instruments, and radios are also made here. The Black Forest is also the setting for many familiar German fairy tales. Select one of these fairy tales and tell why the setting is important to the story.

_____
_____
_____
_____

Name_____

CHINA
from
ENGLAND

England is known for its beautiful china that is decorated with Oriental as well as English designs. This special porcelain dinnerware is known as fine bone china. From what ingredients is bone china made?

_____

_____

_____

_____

_____

Name_____

KLOMPEN
from
THE NETHERLANDS

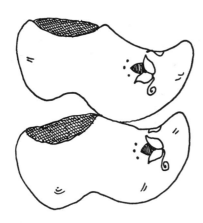

Traditional clothing worn by the Dutch people includes caps, baggy pants, and wooden shoes called klompen. After reading about the land regions in the Netherlands, tell why most Dutch people who live in farm areas and fishing villages prefer to wear klompen instead of leather shoes when they are outdoors.

_____

_____

_____

_____

_____

Name_____

ONCE UPON A TIME . . .

Famous authors such as Jakob and Wilhelm Grimm, Hans Christian Andersen, Sir James M. Barrie, and Charles Perrault wrote many favorite European folk tales and fairy tales. In these stories, good always overcomes evil.

Read a folk tale or fairy tale. Write a different ending for the story.

Name of book: _____

Author: _____

_____

_____

_____

_____

_____

_____

_____

_____

_____

_____

_____

_____

_____

_____

_____

Name _____

LITERATURE SECTION | EUROPE

THE PAINTER'S TRICK
by Piero and Marisa Ventura

COUNTRY: Italy
SUMMARY: A traveling painter tricks a monastery of monks to get food.

New York: Random House, 1977

ACTIVITIES

1. Pretend you have been asked to paint a scene to display in your school's main hallway. What would the subject of your creation be? Create a fresco (painting with watercolors on wet plaster).

2. List the foods mentioned in this book.

    a. Find out what types of livestock the farmers in Italy raise for meat.

    b. What vegetables and fruits are grown in the "boot-shaped" country? Make a product map.

    c. In Europe, cheese is usually made from the milk of goats, sheep, and cows. What types of cheese come from Italy? Have a cheese-tasting party.

    d. Most people in Italy eat yeast breads. Practice your math skills and make a loaf of bread. This is a perfect time to read TONY'S BREAD by Tomie dePaola.

3. Read about famous artists from Italy. Mark their places of birth on a map of the country. Do they represent different regions of Italy?

4. Find out about the famous chapel in which Michelangelo painted frescoes. What is it called and where is it located? How long did it take him to complete the project?

5. Choose one occupation of the Italian people. Write an imaginary paragraph describing a typical "day on the job."

6. Compare this book to *STONE SOUP* told by Marica Brown.

MY UNCLE NIKOS
by Julie Delton

COUNTRY: Greece
SUMMARY: Uncle Nikos, who lives in a little village of Greece, shares his way of life with his niece, Helena, who has come for a summer visit.

New York: Thomas Y. Crowell, 1983

ACTIVITIES

1. Make a list of the Greek words used in this book. Try to discover their meanings through the context of the story.

2. Morning glories grow over the doorway of Uncle Nikos's house. Plant morning glory seeds and record on a graph their daily growth.

3. Rewrite Uncle Nikos's shopping list. Convert the kilo measurement to pounds. Next, take the list to the grocery store and record the price of each item. A comparison can be made if students visit different stores. Approximately how much did Uncle Nikos spend?

4. Small farms are scattered throughout Greece. Make a list of the vegetables and fruits grown on Uncle Nikos's farm. Don't forget the tree crops. You may need to use your research skills to find out whether olives are a fruit or a vegetable. Which of these fruits and vegetables could be grown in your climate? (This would be an excellent cooperative learning project.)

5. Did you know that Greece is a major world producer of raisins? Describe the process of making raisins.

6. The Olympic games originated in ancient Greece. Write a paragraph explaining the purpose of the early competitions. What sports were included?

NORTH AMERICA

The continent of North America is centered between the Atlantic Ocean to the east and the Pacific Ocean to the west. North America intersects with major trade routes of the world and extends from the Arctic Ocean in the north to the country of Panama in the south. It is the third largest continent in size and lies entirely above the equator. North America consists of 23 independent nations. Most of the people speak Spanish, English, or French.

The most northern region of North America is Greenland, the world's largest island. It lies in the North Atlantic Ocean, mainly north of the Arctic Circle. Greenland's small population lives mostly in its warmest region along the southwestern coast. Most of Greenland is covered by a large icecap that always remains frozen. Fishing is a major industry. Weather stations on the island are also important for forecasting storms.

Canada, the second largest country in land area, is located north of the United States. French and English are both official languages in Canada. Most Canadians live in the southern region because of the severe climate in the north. There are large forests and fertile farming areas. The country has rich deposits of copper, silver, gold, uranium, and the largest nickel mines in the world.

The continental United States is located south of Canada and north of Mexico. The state of Alaska is located northwest of Canada. Most United States citizens have ancestors from Europe. English is the chief language. A climate of long summers and warm temperatures

predominates, allowing for extensive farming and raising of livestock. Most of the farmland is located in the midwest. The grasslands of the west are used primarily for ranching. There are also deserts in the west. Forests in the northeast and northwest are rich sources of lumber. Deposits of iron and coal are located in the east. Most factories using these materials are also located there.

Central America is the land to the south of the United States where Spanish is the main language. Most of the people are very poor. Mountains, plateaus, deserts, grasslands, and tropical forests are found in this section of North America. Mexico is the largest country. Its capital, Mexico City, is the largest city in the world. The economy of Mexico is based on agriculture, manufacturing, mining, and tourism. Mexico is famous for its skilled silver craftsmen who frequently display their works in open markets. The people of the other seven Central Middle American countries earn their incomes primarily by working on plantations, in forestry, and by mining. Coffee and bananas are the chief products.

The West Indies, or "Sugar Islands," are a group of islands in the Caribbean Sea formed by volcanic eruptions, limestone, and coral. Most of the people are descendants of black Africans and speak several languages including Spanish and English. Farming is the main occupation. Sugar cane is the main crop. Because of the tropical climate, tourism is the second largest industry.

INDEPENDENT

Antigua & Barbuda
Bahamas
Barbados
Belize
Canada
Costa Rica
Cuba
Dominica
Dominican Republic
El Salvador
Grenada
Guatemala
Haiti
Honduras
Jamaica
Mexico
Nicaragua
Panama
St. Christopher & Nevis
St. Lucia
St. Vincent & the Grenadines
Trinidad & Tobago
United States

DEPENDENT

Anguilla
Aruba
Bermuda
Cayman Islands
Greenland
Guadeloupe
Martinique
Montserrat
Netherlands Antilles
Puerto Rico
St. Pierre and Miquelon
Turks and Caicos Islands
Virgin Islands (U.S.)
Virgin Islands (British)

OUTLINE OF ACTIVITIES | NORTH AMERICA

1. **OUR HERITAGE** *(p. 104)*—The ancestors of most Americans came from other countries. Locate the country or countries from which your relatives came. Select one of the countries, and dress the paper character in the traditional costume of the country. List five interesting facts about the country on the back of the character. Be sure to include the chief products, largest city, and one place of interest to visit.

2. **CANADIAN WILDLIFE** *(p. 105)*—Canada is divided into ten provinces. Select one of the provinces and draw a large map of it on a posterboard. On the map, draw some of the birds or mammals that inhabit this area. Next, select one of the birds or mammals that interest you and write a paragraph persuading your community to protect its existence.

3. **TAKE A TRIP THROUGH THE PANAMA CANAL** *(pp. 106-107)*—You have decided to take a trip from New York City, New York, to San Francisco, California, by **BOAT**. One route would take you around South America, while the other is through the Panama Canal, one of the greatest engineering achievements in the world. Use a map to determine which route to take. Defend your reasons for making this choice.

4. **YOU'RE INVITED TO A FIESTA** *(pp. 108-109)*—In Mexico, the people love fiestas. Plan a fiesta. Design an appropriate invitation. Write a paragraph explaining how to plan a fiesta. Don't forget to include the food and activities that will be enjoyed at the fiesta.

5. **IT'S A HURRICANE** *(p. 110)*—Hurricanes are powerful, destructive storms that normally occur in tropical regions. Find out more about these storms. Make a list of preparations for your family.

MORE TOPICS TO RESEARCH

Panama Canal	Alamo	North Atlantic
Maple syrup	Niagara Falls	icebergs
production	Grand Canyon	Sugar cane
Salmon industry	Alaskan Highway	Totem poles
Inuit	Lacrosse	Quetzal
Paper production	Royal Canadian	Musk ox
Scrimshaw	Mounted Police	Sapodilla
Bald eagle	Alaskan pipeline	Olmec Indians
Northwest Passage	Mass production	
Adobe	Cape Canaveral	

The ancestors of most Americans came from other countries. Locate the country or countries from which your relatives came, select one country, and use paper and/or cloth to dress the paper character in the traditional dress of that country. List five interesting facts about the country on the back of the character.

1. _____
_____
_____

2. _____
_____
_____

3. _____
_____
_____

4. _____
_____
_____

5. _____
_____
_____

CANADIAN WILDLIFE | NORTH AMERICA

Canada is divided into ten provinces. Select one of the provinces and draw a large map of it on a posterboard. On the map, draw some of either the birds or mammals that inhabit this area. Next, select one of the birds or mammals that interest you and write a paragraph persuading your community to protect its existence.

TAKE A TRIP THROUGH . . .

You have decided to take a trip from New York City, New York, to San Francisco, California, by BOAT. There are two routes that you can take. One route would lead you around the southern tip of South America, while the other runs through the Panama Canal, one of the greatest engineering achievements in the world.

How many miles is the trip from New York to San Francisco through the Panama Canal?

_____

How many miles is the trip from New York to San Francisco around South America?

_____

Choose one of these routes for your trip. Give three reasons why you believe this would be the better route to take.

1. _____

_____

_____

_____

2. _____

_____

_____

_____

3. _____

_____

_____

Name _____

© 1993 by Incentive Publications, Inc., Nashville, TN.

Two types of boats are too large to travel through the canal. Can you name them?

1. _____

2. _____

Design a boat on which you could travel through the canal.

As you pass through the canal region, you may see large plantations where Panama's chief CASH crops are grown. What are cash crops?

_____

_____

_____

Name two types of cash crops grown in Panama.

1. _____

2. _____

Name _____

You're Invited To A

The Mexican people love fiestas. These festivals or parties celebrate national or religious holidays. It's your turn to plan a fiesta!

Cut out the invitation below and fold it in half. Decorate the front to represent a specific holiday. You will need to research the dates on which particular holidays are observed in Mexico. Fill in the information on the invitation about the party.

fold

Date: _____

To celebrate: _____

_____

Time: _____

Place: _____

Given by: _____

fold

FIESTA!!!

Fill in the lines below.
Then, on a separate
sheet of paper, write a
paragraph explaining
how to plan a fiesta.
Include as many
details as possible.
Don't forget to mention the holiday you are
celebrating.

Foods to serve:

1. _____

2. _____

3. _____

4. _____

5. _____

Activities to enjoy:

1. _____

_____

2. _____

_____

3. _____

_____

4. _____

_____

Name _____

IT'S A HURRICANE!

Hurricanes are powerful, destructive storms that normally occur in tropical regions. In North America, these storms usually strike the islands of the West Indies and land areas along the Gulf of Mexico and Atlantic coastlines.

1. During what months do most hurricanes occur? _____

2. Find the wind speeds for each of the following types of storms:

    Tropical depression _____ m.p.h.

    Tropical storm _____ m.p.h.

    Hurricane _____ m.p.h.

3. On a separate sheet of paper, make a list of supplies that you should have available if a storm should cross your area. REMEMBER—You may not have electricity!

4. Make a list of precautions you should take and things you should do to protect your family and your property.

5. You are a meteorologist at the National Weather Service. Write a news bulletin describing a hurricane. Include information about how hurricanes form.

6. If possible, use a hurricane tracking chart to track a hurricane.

7. Draw a picture of the warning flags that you would find flying on the coastline to alert ships of a hurricane.

Name _____

MY GRANDPA AND THE SEA
by Katherine Orr

COUNTRY: St. Lucia
SUMMARY: Grandpa is forced to find a different means of supporting his family because the larger boats have depleted the island's supply of fish.

Minneapolis, MN: Carolrhoda Books, 1990.

ACTIVITIES

1. Make a mural of the different occupations on the island of St. Lucia.

2. Compare a dugout to a modern fishing boat.

3. Explain what Grandpa meant when he said, "If we give back something for everything we take, we will always meet with abundance."

4. Ecologists are concerned that man is depleting valuable resources. Make a list of these resources and brainstorm ways to save them.

5. Find out more about the occupation of a marine ecologist.

6. Use the following question to conduct a class debate: Are our lives being strengthened or weakened by modern technology?

7. Read the book *MY LITTLE ISLAND* by Frane Lessac. Make a list of some of the customs on the Caribbean island as described in the book. The author does not identify the island by name. Using the context clues and your research skills, try to discover which island is being described.

MOLLY'S PILGRIM
by Barbara Cohen

COUNTRY: United States
SUMMARY: A young Russian immigrant, Molly, must accept the cultural changes in her life, and, at the same time, try to make her new friends understand her heritage and culture.

New York: Lothrop, Lee and Shepard, 1983.

ACTIVITIES

1. Explore your family's past and discover your country of origin. On a large world map, mark the country with your name.

2. Discuss the reasons people immigrated to America. Use the pilgrims as one example.

3. Dress a heritage character (activity #1 - North America) to represent the traditional clothing of your country of origin.

4. As a class project, build a three-dimensional early American settlement and display the heritage characters. Include a Native American village complete with tepees, canoes, campfires, and painted animal skins.

Use hieroglyphics to tell a story on animal skins. First cut a brown paper sack in the shape of an animal skin. Draw your story in hieroglyphics on the shape. Crinkle the shape into a ball. Then spread it out again.

5. Design a family shield. Explain the reason you chose each symbol for the shield.

6. What preconceived ideas did Molly's class have about Russians? Make a list of what the children learned from Molly about her culture, and what Molly learned from the children about America.

7. Use the encyclopedia to find out about Ellis Island.

SOUTH AMERICA

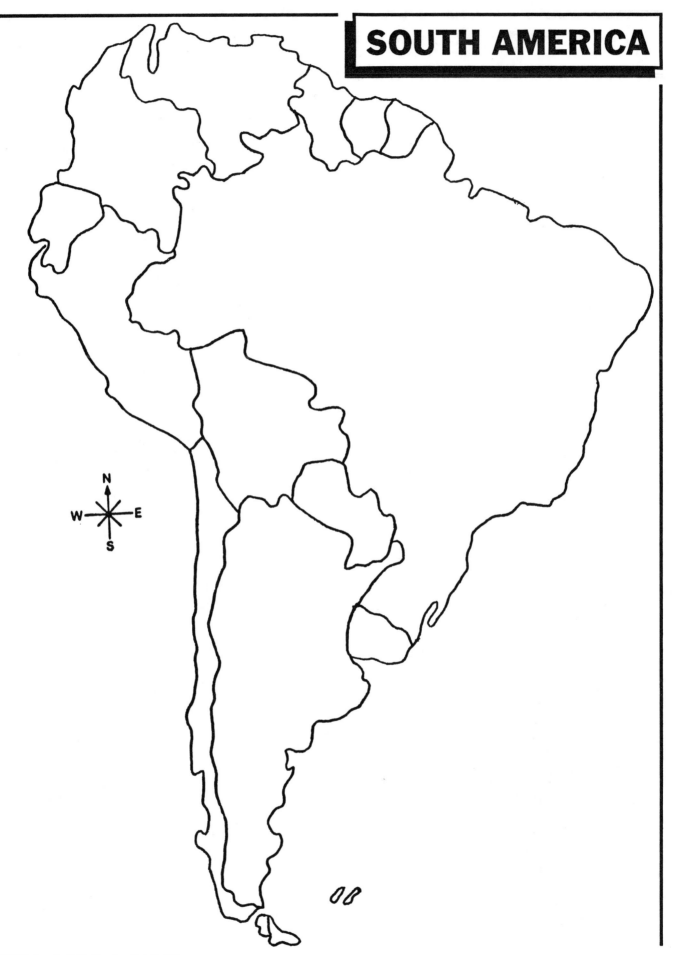

South America is the fourth-largest continent. It is surrounded mostly by water. Only a narrow strip of land connects the continents of South America and North America. The Caribbean Sea is located to the north of the continent. The Atlantic Ocean lies on the east, the Pacific Ocean lies on the west, and the Drake Passage separates South America from Antarctica on the south. The landscape of this continent, which consists of 12 independent nations, includes the snow-peaked mountains of the Andes, and one of the world's driest deserts, the Atacama. Most of the people live near the coastlines. The fishing industry is very important along the coasts of Chile and Peru. Spanish is the official language of South America; however, Portuguese is spoken in Brazil. The countries of Argentina, Brazil, Uruguay, and Venezuela have the most highly developed economies. Three main land regions cover the continent.

The Andes Mountains, the longest visible mountain range in the world, cover the west coast of South America from Venezuela, a petroleum-rich country, to the southern tip of the continent. Mineral deposits of copper, gold, tin, and zinc are found in the mountains. The country of Bolivia leads the world in tin mining, while Colombia produces 90% of the world's emeralds. The fertile land of the mountain slopes is also used to grow crops such as coffee beans. The region is also used to graze sheep, cattle, and llama in the countries of Peru and Chile.

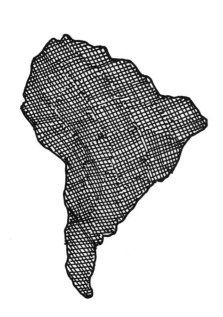

OVERVIEW | SOUTH AMERICA

The Central Plains cover almost half of the continent. They project eastward from the Andes Mountains. Grasslands called llanos cover the land in Venezuela and Colombia. Some of the largest farms in the world are located in this area. The selva, or tropical rain forest, is also located in the Central Plains. The Amazon River basin countries of Brazil, Peru, and Bolivia are located in this area. The natural environment of this area is endangered because so many of the trees are being cut down for forest products. Forest resources such as rubber trees, rosewood, mahogany, and balsa wood are also found in the Gran Chaco area. Vast grassland areas called the pampa are found in the southern regions of Argentina. This area is good for sheep and cattle ranching. The hottest point in South America is located in this region.

The Eastern Highlands include the mountain areas to the north and south of the Amazon River. Tropical rain forests and grasslands are found in the highlands north of the Amazon. Rolling hills and flat land are located to the south. This is rich farmland. Much of the coffee grown in Brazil, "the world's coffeepot," comes from this area.

Unusual animals found in South America include the manatee, the capybara, the toucan, the flamingo, the vicuna, and the llama.

COUNTRIES

INDEPENDENT

Argentina
Bolivia
Brazil
Chile
Colombia
Ecuador
Guyana
Paraguay
Peru
Suriname
Uruguay
Venezuela

DEPENDENT

Falkland Islands
French Guiana

OUTLINE OF ACTIVITIES | SOUTH AMERICA

1. **SCHOOL DAYS, SCHOOL DAYS** *(p. 118)*—Using two gelatin boxes, make a puppet dressed as a child from a country in South America. Prepare a monologue for your puppet about education in that country.

2. **FACT-FINDING FESTIVAL**—Using the encyclopedia, locate an article about one of the countries in South America. Use the subtopics in that article to list facts about each of the subjects listed below.
   - I. Brief Facts
     - A. The capital city
     - B. The population
     - C. The chief products
     - D. The basic unit of money
   - II. The People
   - III. The Land (include weather information)
   - IV. The Government
   - V. Occupations

3. **HAVING A GREAT TIME** *(p. 121)*—After completing the shape booklet (#2), pretend that you are visiting the country. Write and address a postcard to a friend, stating at least three facts about the country that you are visiting. Draw a scene of the country on the back.

4. **FUN IN THE SUN** *(p. 122)*—Using a box lid for your display, design a billboard advertising a form of recreation in a South American country.

5. **PRODUCTS OF SOUTH AMERICA** *(p. 123)*—On an outline map of South America, draw the products grown on that continent. Be sure to draw the products in the areas where they are grown. Include a pictorial legend to identify the products.

6. **ANIMALS, ANIMALS, EVERYWHERE** *(p. 124)*—From the 500 section of the library, select an animal book. After reading the book, read about the same animal in the encyclopedia. List three facts you learned about the animal from the encyclopedia article that were not in the book.

MORE TOPICS TO RESEARCH

Monkey Puzzle tree	Easter Island	Manatee
Bullfighting	Cape Horn	Tapir
Gaucho	Cinchona tree	Forest products of
Pan American	Cocoa tree	Brazil
Highway	Vicuña	Brazilwood
Pelé	Coffee bean	Andes Mountains
Llama	processing	Sugar Loaf Mountain
Rubber production	Strait of Magellan	Tierra del Fuego
Amazon River	Inca Indians	
Angel Falls	Lake Titicaca	

SOUTH AMERICA

SCHOOL DAYS, SCHOOL DAYS

The educational requirements of young people are different in other countries. Select one of the countries in South America, and read about the education of its children. Using two gelatin boxes, make a puppet dressed in the traditional clothing of that country. Prepare a monologue for your puppet about the education in the country, and present your findings to your class.

HOW TO MAKE THE PUPPET:

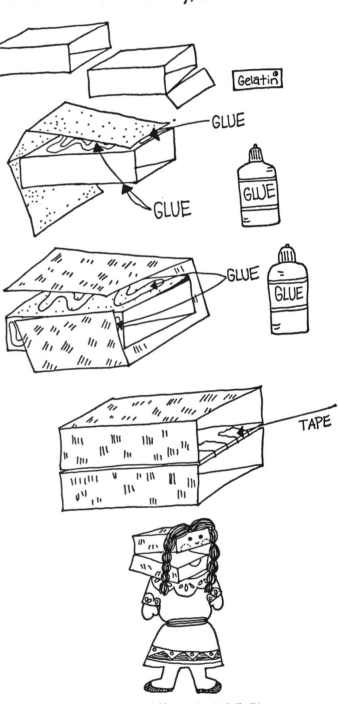

1. Take 2 small gelatin boxes, and cut off the flaps of one end of each box.

2. Apply glue to the box. Using a strip of paper 2¾" wide by 7¾" long, wrap the box from the top edge of the open end, over the box, and continuing to the bottom of the open end. Leave the open end uncovered.

3. Use a strip of paper 3¼" wide by 6½" long to cover the remaining sides of the box.

4. Repeat steps #2 and #3 on the second box.

5. Hinge the two boxes together using tape on the open edges.

6. Decorate the head.

7. Finally, design a body dressed in traditional clothing, and tape it under the head.

FACT-FINDING FESTIVAL | SOUTH AMERICA

Long ago the people of South America made beautiful masks from thin sheets of gold. The masks were worn during ceremonies and festivals. Compile a "shape" booklet of facts about one of the countries of South America. Cut the pages in the shape of the mask pattern on page 120. Staple the pages together. Using the encyclopedia, locate an article about the country. List facts about each of the topics listed below. Use one page of the booklet for each topic. Cut out the cover from the same mask-shaped pattern. Decorate the cover as described on page 120. Use pieces of colored paper, feathers, or other materials to decorate the cover to resemble an ancient mask.

I. Brief facts
   A. The capital city
   B. The population
   C. The chief products
   D. The basic unit of money
II. The people
III. The land (include weather information)
IV. The government
V. Occupations

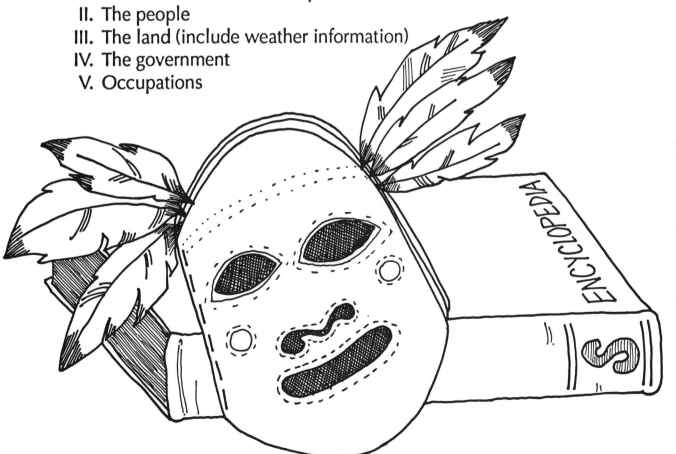

After completing the shape booklet (pages 119-120), pretend that you are visiting the selected country. Write and address a postcard to a friend, stating at least three facts about the country that you are visiting. Draw a scene of the country on the back of the postcard. Design a stamp that would represent a species of animal from that country.

Dear _____ ,

Your friend,

_____

SOUTH AMERICAN TAPIR

SOUTH AMERICA | FUN IN THE SUN

People in most countries enjoy having fun during their leisure time. Some people prefer to participate in various types of outdoor activities, while others choose to be spectators of different events.

Select a South American country. Read about its recreational activities, sports, and ways of life. Sometimes the climate and land regions are clues to the different outdoor events that take place in the country.

Using a small box lid for your display, design a billboard to advertise a form of recreation in the South American country that you chose. Exhibit the billboard in your classroom.

PRODUCTS OF SOUTH AMERICA

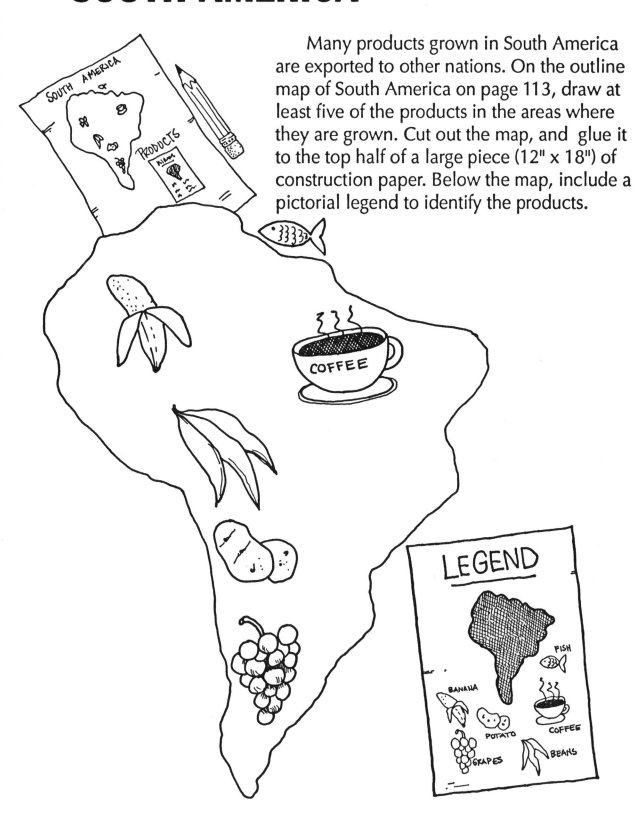

Many products grown in South America are exported to other nations. On the outline map of South America on page 113, draw at least five of the products in the areas where they are grown. Cut out the map, and glue it to the top half of a large piece (12" x 18") of construction paper. Below the map, include a pictorial legend to identify the products.

SOUTH AMERICA

ANIMALS, ANIMALS, EVERYWHERE

Although there are many types of animals, most require specific conditions in order to survive. Climate and geography are two factors that play a key role in forming an animal's habitat.

Select a book about any animal. After reading the book, read about the same animal in the encyclopedia. List three facts that you learned about the animal from the encyclopedia article that *WERE NOT* in the book.

Name of book: _____

Author: _____

Type of animal: _____

This animal would most likely

be found on the continent of _____.

Additional facts learned about this animal from the ENCYCLOPEDIA.

1. _____

   _____

2. _____

   _____

3. _____

   _____

Name _____

THE GREAT KAPOK TREE:
A TALE OF THE AMAZON RAIN FOREST
by Lynne Cherry

COUNTRY: Brazil
SUMMARY: The animals of the tropical rain forest in Brazil try to persuade a man not to destroy their homes by cutting down the trees.

New York: Gulliver Books, 1990

ACTIVITIES

1. List some animals that live in tropical rain forests. Research each animal and combine the information to make a class book.

2. Outline the reasons the animals gave for saving the trees. Select one of the animals and its reason for saving the forest. Convince the class that this animal has the BEST reason of all.

3. Read about the kapok tree.

4. Brainstorm to answer this question: Why is it important to save the rain forest?

5. Find out what governments around the world are doing to save the rain forests.

6. List products that come from the trees of the rain forest. Can these products be recycled in order to save the trees?

7. Locate the tropical rain forests around the world and mark them on a map. Estimate the percentage of rain forests that have been destroyed. The end pages in *The Great Kapok Tree* can be used to determine this information.

LLAMA AND THE GREAT FLOOD:
A FOLKTALE FROM PERU
by Ellen Alexander

COUNTRY: Peru
SUMMARY: The llama warns his master of a forthcoming flood and saves the family by taking them to a high peak in the Andes Mountains.

New York: Thomas Y. Crowell, 1989

ACTIVITIES

1. Find out more about the llama. List at least five ways this animal is useful to the Indians who live in the Andes Mountains.

2. The Inca Indians, as pictured in the book, produced beautiful crafts. Most of these crafts were used in their day-to-day life. Make an example of an Inca craft and tell how the Incas would have used it. (Cardboard looms can be used for weaving.)

3. The Andes Mountains are the longest chain of mountains above sea level in the world. They include many land forms, natural resources, and animals. Travel through the mountains is often difficult. Research one of the following topics as it relates to the Andes and write a short paragraph. Make a class booklet about the Andes Mountains.

a. glaciers
b. lakes
c. rivers
d. natural resources
e. types of transportation
f. alpaca
g. chinchilla
h. condor
i. huemul
j. llama

4. The Incas tied colorful yarn containing special herbs to the ears of the llama. They believed that this would keep the animal healthy and protect its owner. Although there is no written record of this, perhaps the ties were also used for identification. List animals on which we place markings of identification. Describe the methods used to mark the animals.

5. Read more about the present-day Indians of Peru. Compare them to their ancestors.

CONCLUDING YOUR TRAVELS . . .

Plan an international celebration day to culminate your study. The students will look forward to sharing what they have learned. They will also enjoy observing and participating in demonstrations of different cultures.

1. Organize a parade to begin the day.
   a. Ask students to dress in costumes from the countries located on a specific continent.
   b. Use wagons to design floats that depict the different countries on the continent.
   c. Allow the students to carry handmade flags, as well as artifacts and stuffed animals, that represent the different countries.
   d. Play ethnic music during the parade to set the mood.

2. Invite parents and other students to visit your room. During this time your students can share what they have learned.

3. The day is extra special if each grade level in your school has studied a particular continent. Decorate each grade level to represent a certain continent, with each room of that grade level representing a country on the continent.
   a. Assign parents to each room to cook ethnic foods with the class, demonstrate crafts, teach games, sing songs, and explain different artifacts from around the continent.
   b. This is also a perfect opportunity to invite guest speakers, dancers, storytellers, etc.

4. Encourage parents and special guests to dress in authentic costumes.

5. Plan for the children to move from room to room throughout the day, "traveling around the world."